Laboratory Manual for

Electronics via Waveform Analysis

Edwin C. Craig

Laboratory Manual for
Electronics via
Waveform Analysis

With 36 Figures

Springer-Verlag
New York Berlin Heidelberg London Paris
Tokyo Hong Kong Barcelona Budapest

Edwin C. Craig
7660 South Wilbur Wright Road
Cambridge City, IN 47327
USA

© 1994 Springer-Verlag New York, Inc.
Reprint of the original edition 1994

Production managed by Francine McNeill; manufacturing supervised by Vincent Scelta.
Camera-ready copy supplied by the author using Microsoft Word.

9 8 7 6 5 4 3 2 1

ISBN-13:978-0-387-94136-3   e-ISBN-13:978-1-4612-2610-9
DOI: 10.1007/978-1-4612-2610-9

Table of Contents

Introduction

To the Instructor

The purpose of this laboratory manual is not just to help students to set up electronic circuits that function as they should. The important thing is the electronic concepts that the student learns in the process of setting up and studying these circuits. Quite often a student learns more electronics when he has to trouble shoot a circuit than when the circuit performs as it should when first built.

It is unlikely that any students would be able to complete all of these experiments in one semester. The author believes that all students should have laboratory experiences with power supplies, amplifiers, oscillators, and integrated circuits. Additional laboratory experiments should be determined by the instructor. Therefore, you can choose those that you want done. Some students are more efficient in the laboratory than others. Therefore, some would be able to complete more experiments in a semester than others. Also many of these experiments cannot be completed in one two-hour laboratory period. If space is available, the circuits could be left intact from one period to the next. Or you might want to select steps in an experiment that you want to delete. Neither the values of the components or the magnitudes of the power supplies, as given in the instructions, are critical. Therefore you could in most cases change them if the ones recommended are not available.

Many instructors differ in the type of reports that they want the students to turn in along with their data at the end of laboratory experiments. Tables have been included in most of the experiments to help the students in reporting their data. In some of these experiments the author has made suggestions concerning the report. In other cases questions were asked along with the instructions at different steps in the procedure. It is up to the instructor to determine the type of report that is desired for each experiment.

To the Student

Do not forget that the purpose of performing experiments in the laboratory is to reinforce or add to the knowledge that is learned in the classroom. Too many students in an electronics laboratory are concerned only in setting up an electronic circuit that functions as it should. This is important, but it is more important that various tests and/or studies be made with the circuit after it is built.

Neither the values of the circuit components or the magnitudes of the power supplies are critical in most of these experiments. If you do not have what has been suggested in the instructions, then use a component or power source that is reasonably close to the suggested value. As an example, most transistors and integrated circuits can operate with power supplies from 9 to 20 V and sometimes more. Do not become frustrated if a circuit does not function properly when first constructed. When the author would find a student whose circuit was not functioning as it should the student was told, "Good, now you can learn some electronics by trouble shooting the circuit."

After you have completed the suggested procedures in an experiment, you might want to perform other tests on the circuit. It is not absolutely necessary that every part of every experiment be performed. If too much time is being spent on an experiment ask your instructor if some parts can be deleted. Neither is it necessary that every experiment in this manual be performed in a semester. The selection of those to be performed should be determined by your instructor.

Major Equipment Needed

1. Oscilloscope with following capabilities: Triggered sweep, dual beam, invert one channel, and sum of channels one and two. If your oscilloscope does not have the last two capabilities two simple circuits, that perform these two functions, are included at the end of this list of equipment.
2. VOM (volt-ohm-milliammeter): With ac voltage ranges, high impedance, for example, FET or digital.
3. Function generator: These may be purchased in kit form. These kits are inexpensive and knowledge of electronics is not needed to follow the instructions provided with the kits. These are sometimes called waveform generators. Sine wave oscillators can be used but they are not as good as function generators. The function generators provide square waves and triangular waves, in addition to sine waves. The square waves are very useful as external trigger voltages for the oscilloscope.
4. Resistance substitution box: This may also be purchased in kit form which is easy to construct.
5. Solderless breadboard.
6. A dc power supply: With outputs of both +15 and −15 V. Two 9-V transistor batteries could be used in series, when needed, instead of a rectified power supply.

The following items could be shared by lab groups:
1. One semiconductor specifications manual.
2. One or more soldering irons.
3. One roll of 60/40 solder 0.031 diameter.
4. A small supply of cloth or plastic covered No. 22 to 29 solid copper wire. The diameters of the leads on some components are too large to use with solderless breadboards, so small sections of the above wire needs to be soldered on these leads.
5. A small wire stripper.
6. Long nose pliers.
7. A supply of alligator clips, spring clips, or other components, needed to connect the components together in those experiments in which the solderless breadboards are not used.
8. A supply of standard value 1/4-W carbon resistors. The leads on 1/4-W resistors are of the proper size to insert in the solderless breadboards. Most companies have kits of resistors that contain the standard values needed for the experiments. Some catalogs show these standard values. In most experiments the resistance values are not critical so that resistors with 10%, or

even 20%, tolerance are suitable. A list of some of the most used standard values of resistances is included in Table I.1.

9. Capacitors: A small supply of 0.1- and 0.01-μF, and 220- and 470-pF capacitors (not electrolytic), and 47- and 100-μF electrolytic capacitors.

10. The special components needed for each experiment will be listed at the beginning of the instructions for each exercise.

Table I.1. A list of standard values of resistance, in ohms, plus a few other useful values.

47	680	4.7 k	33 k	220 k
68	1.0 k	6.8 k	47 k	330 k
100	1.5 k	10 k	50 k	360 k
270	2.2 k	20 k	68 k	680 k
330	2.7 k	22 k	82 k	1 M
470	3.3 k	27 k	100 k	3.3 M
560	3.9 k	30 k	150 k	4.7 M

The following two circuits require a minimum of components and can be built on solderless breadboards and left intact for use in any of the experiments. A 9-V transistor battery can be used as the power supply for each circuit.

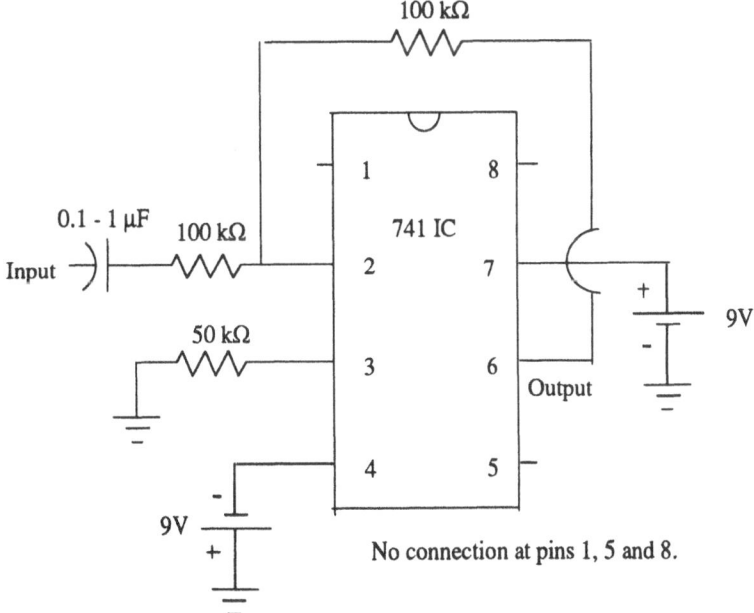

Figure I.1. Circuit that can be used to invert a signal without affecting its amplitude.

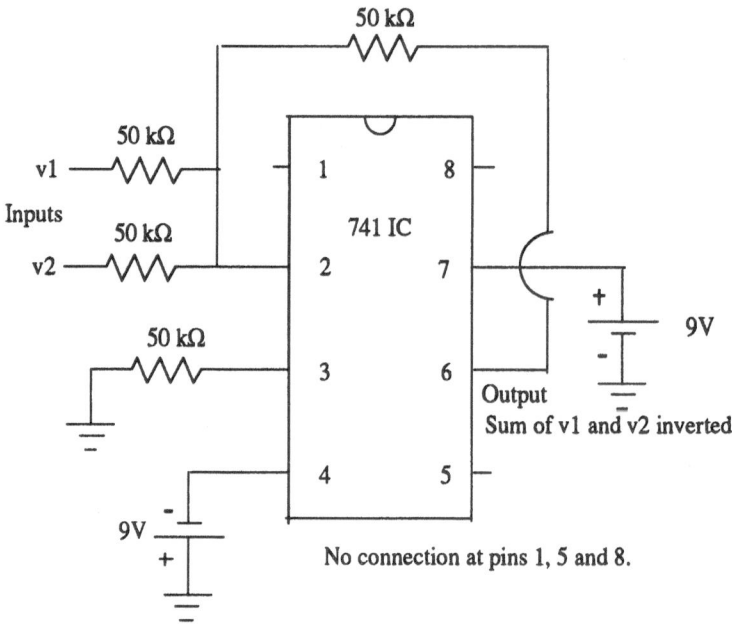

Figure I.2. Circuit that can be used to add two voltage waveforms.

1
Half-Wave Power Supplies

Components Needed

1. Transformer: 115 V primary and a center-tapped secondary, with a full secondary voltage of less than 20 V.
2. Diodes: Four silicon rectifier. Only one will be needed in the half–wave power supply.
3. Capacitors: Three electrolytic with voltage ratings of 50 V or more, one of 50 μF or less, and two of 200 μF or more.
4. Resistors: Two of any value from 1 to 20 kΩ and power ratings of 5 W (watts) or more.
5. Alligator clips: Ten or more spring loaded clips, or other devices to connect the components in the circuit. The leads on the components used in these experiments have diameters too large to use with solderless breadboards.

Before starting to build any of the rectifier circuits, be sure to check the silicon diodes to make sure that they are not defective. Also determine which end of each diode is the anode and which is the cathode. When a VOM meter is being used as an ohmmeter, one of the test leads has a positive dc voltage with respect to the other and it is necessary to determine which one it is. In order to determine this, connect a second dc voltmeter to these test leads and observe the polarity on this second voltmeter. When testing your silicon diodes with an ohmmeter set the multimeter to the ohms ×10 scale and zero it with the leads shorted together. Now clip one test lead to one end of a diode and then touch the other test lead to the other end of the diode. Do not clip this second lead to the diode. If the meter registers a low resistance (less than 20 Ω) then the end of the diode that is connected to the positive lead of the ohmmeter is the anode of the diode. This is the end opposite the arrow in the symbol (▸⊦). Now reverse the connections of the ohmmeter to the diode. If the diode is not defective the meter should register a high resistance in this direction. The two resistances should have a ratio of at least 100 to 1 for a good diode.

Important

In each experiment that you do, when you display a waveform on the oscilloscope, take the time to analyze that waveform and relate it to the circuit diagram. Ask yourself, "what is happening in the circuit to produce the waveform ? "

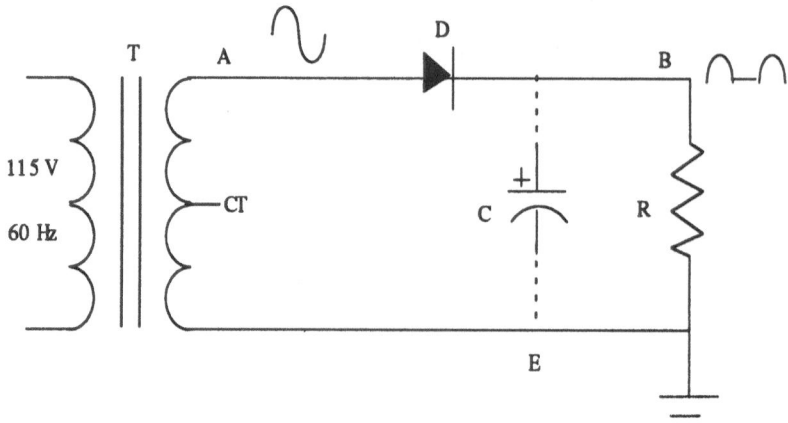

Figure 1.1. Half–wave rectifier circuit.

If there are any distortions in the waveform try to determine the cause. Be on the lookout for small dips or bumps on the waveforms. In this way you will become more knowledgeable in electronics. It may help if you refer back to the discussion of the circuit waveforms in the text from time to time. Compare your voltage waveforms to those photographed and included in the text. If an oscilloscope camera is available, photographs of the waveforms could be used instead of the drawings in the tables, and other parts of the reports for the experiments.

It is very important that you obtain external trigger voltage from the same point in the circuit during a complete experiment. If that is done, then all of the points in the different waveforms, that are on the same vertical line on the oscilloscope, occur at the same instant in the circuit being studied.

Half-Wave Rectifiers

Build the circuit of Fig. 1.1 without the capacitor C in the circuit for the first part of the experiment. Use a 1- to10-kΩ resistor with a power rating of 5 W or more for R.

1.

Using a multimeter, measure the ac voltage at point A with respect to the common point E.
$v_A =$ _____ volts rms.

2.

With the oscilloscope switched to "external trigger, " use point A to get the trigger voltage for the entire experiment. Connect the ground lead of the oscilloscope to point E.

Now, observe simultaneously, and measure the amplitudes and durations of the waveforms at points A and B, and record in Table 1.1. Always measure the amplitudes of ac voltages, pulses, etc., from peak to peak unless told to do otherwise in the instructions. Make the drawings of two cycles of the waveforms in the same relative positions in the x direction (horizontal) as they would appear on the oscilloscope when displayed simultaneously.

Why does a half sine wave at B appear directly above only the positive half of the sine wave at A and not above the negative half?

3.

Now, zero center the sine wave at point A on a grid line on the oscilloscope by either temporarily moving the ac-ground-dc switch to "ground," if there is one on your oscilloscope, or by temporarily disconnecting your signal lead from the circuit, and then moving the horizontal line so that it is on one of the lines on the grid of the oscilloscope. If this horizontal line is sinusoidal instead of straight, when the signal probe is disconnected from the circuit, this is due to 60 cycle pickup by the signal probe. In order to get rid of this, clip the ground lead of the probe to the signal lead. Now observe the waveform at point A. It should be centered on the zero volts line.

4.

Now, find the zero volts line for the probe connected at point B in the same manner as for A above and superimpose this zero volts line on the same grid line as before. Now, when you observe the waveform at point B it should be superimposed on the top half of the waveform at point A. Does there appear to be much voltage loss from point A to point B in the top half of the sine wave? Has this half of the sine wave been distorted by the diode?

Function of Filter Capacitors

Disconnect the ac power from the circuit while connecting the smaller of your electrolytic capacitors in the circuit.

Table 1.1. Data on a half-wave rectifier without C in the circuit.

Point	Two cycles of waveform	Amplitude in V peak-to-peak	Duration of one cycle in ms
A			
B			

Table 1.2. Half-wave rectifier parameters with a small capacitance for C in Fig.1.1.

Point	Two cycles of waveform	Amplitude peak-to-peak	Duration			
			Positive slope		Negative slope	
B		V		ms		ms
			Positive peak		Negative peak	
A		V		ms		ms
dc voltage at point B as measured with the oscilloscope = V						
dc voltage at point B as measured with a multimeter = V						

1.

Now, with C in the circuit, simultaneously observe the voltage at points A and B. You can use either dc or ac in this observation. If the waveform at A and/or B has its positive peak flattened, or distorted, turn off the power to the circuit and measure the resistance of the secondary of the transformer. In the transformer, used by the author to get the photographs in the textbook, the secondary had a resistance of less than 1 Ω. If this resistance for your transformer is much greater than 1 Ω, then there might be an appreciable voltage drop across the secondary of the transformer during the time that the diode is forward biased and the capacitor is being recharged. This could cause distortion of the waveform at point A during the positive ramp at point B.

Move the waveform at B vertically so that it is just above the positive peak at A. What is occurring in the circuit during the time that the positive slope of the ramp is at B? The negative slope? Why is the duration of the negative slope so much greater than that of the positive slope?

2.

Now move the trace at B until the positive ramp sets on the waveform at A. At about what voltage does the diode start conducting current? Note the voltage at point A when the diode is cut off.

3.

Move the dc zero volts line for the waveform at B to an upper grid line, but not the top one, and measure and record the items requested in Table 1.2.

In order to measure dc with the oscilloscope, proceed as follows. If the oscilloscope has a switch marked "ac-dc-ground" then switch it to "ground." A horizontal straight line will appear on the screen. Move this line up or down until it is on one grid line of the oscilloscope screen and then switch the input from ground to dc. Then measure from this grid line to the trace representing the dc voltage at the point at which the probe is connected. If your oscilloscope has no switch that is labeled "ac-ground-dc," it will be necessary to disconnect the probe from the circuit, switch the input to dc, and connect the signal probe to ground. Then position the horizontal line to the grid line on the screen. Then connect the signal probe to the point on the circuit that you want to measure the dc voltage.

The ripple or sawtooth voltage, found at point B, is an ac voltage that is present along with the desired dc voltage in all power supplies that use rectified ac voltages. A good power supply has a ripple with very small amplitude. The positive slope in this sawtooth shows the increase in voltage across the capacitance as it is being charged by current through the diode during the time that it is forward-biased. The long negative ramp shows the voltage across the capacitance as it discharges through R during the time that the diode is reverse-biased.

Effect of Magnitude of Capacitance in C

Replace the smaller capacitor with one that has significantly larger value of capacitance and then make the measurements necessary to complete Table 1.3. Some of the items can be copied from the previous table.

Voltage Drop Across the Diode

Now, let us try to determine the voltage drop across the silicon diode while it is conducting. Use the larger value of C while doing it.

1.

Either switch both channels on your oscilloscope to ground, or disconnect both leads from the circuit. Position both horizontal lines so that they are superimposed at the lowest position possible on the oscilloscope. Using point E as the reference point (ground lead at that point) connect both signal leads to the same point, B. Switch both inputs of the oscilloscope to dc and adjust the vertical gain controls to the same setting so that the horizontal traces are in the upper half of the screen but below the top line. If the oscilloscope is calibrated properly the traces will still be superimposed. If they are then move one signal lead from point B on one side of the diode to point A on the other side. Measure the difference in volts between the trace at point B and the positive peak of the waveform at point A, and record that voltage in the space below.

Voltage drop across this silicon diode was ___ V.

In the step above, if the traces are not superimposed when both signal leads are connected at point B, then at least one channel of your oscilloscope is not calibrated perfectly. However, your results will be reasonably accurate if you move one of the traces until they appear as one line.

Table 1.3. Effect of filter capacitance magnitudes on ripple voltage in dc power supplies.

Point	dc voltage		Amplitude of ripple voltage	
	Small C	Large C	Small C	Large C
B	V	V	V	V

Then take the signal lead that goes with the trace that you did not move and transfer it to point *A*, and record the difference in voltage between the two traces at the peak of the trace at point *A* in the space on p. 9. The accuracy of your result will depend on the calibration of your oscilloscope and your ability to interpolate between the lines as you read the voltage.

Power Supplies that Provide Negative dc Output Voltages

Reverse the diode and remove the capacitor.

1.

Without *C* in the circuit observe the voltage waveforms at points *A* and *B* simultaneously. Draw the waveform at point *B* under two complete cycles of the ac input at *A* in Table 1.4.

2.

Connect the larger value of *C* in the circuit. The polarities should be reversed from those in the positive dc power supply.

3.

Observe the waveforms at points *A* and *B* simultaneously. Draw these waveforms at *A* and *B* in Table 1.4.

4.

Measure the dc voltage at the output of the rectifier (*B*) and record in Table 1.4.

Questions Concerning Half-Wave dc Power Supplies

1. What are the functions of the diode, capacitor, and resistor in the circuit?
2. How does the capacitance perform its function?
3. Describe the action in the circuit that produces the waveform at point B under the following conditions:
a. Without *C* in the circuit.   b. With *C* in the circuit.

Table 1.4. Data on a half-wave rectifier that has a negative dc output.

Point	Voltage waveform without C	Voltage waveform with C
A		
B		

The output voltage of this rectifier was   V.

2
Full–Wave Power Supplies

The components needed for this experiment are the same as for the half–wave rectifier but two silicon diodes will be needed and the center tap of the transformer will be used.

Full–Wave Rectifiers

Build the circuit in Fig. 2.1 without the capacitance C in the circuit. Connect one ground lead of the oscilloscope to point F which is to be used as chassis ground during this experiment. Note that this common point is connected directly to the center tap on the transformer T. Use the external trigger mode and get the trigger voltage at point A.

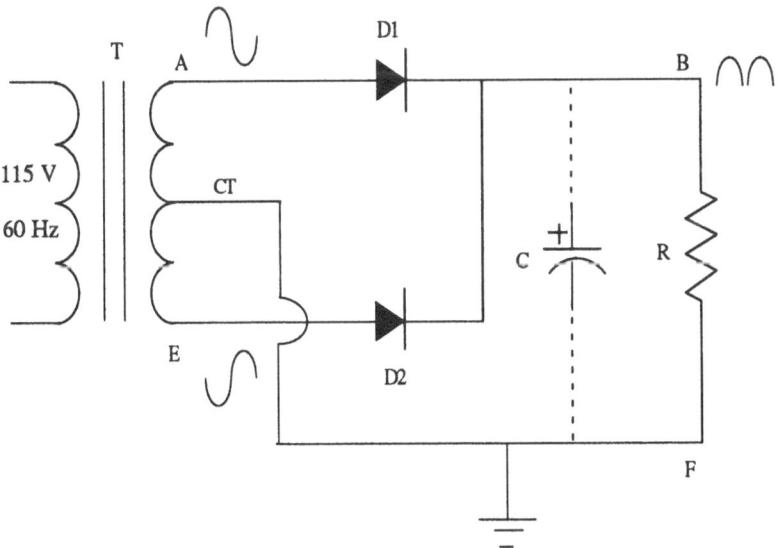

Figure 2.1. Full-wave rectifier using a center–tapped transformer.

1.

Observe the voltage waveforms at the two ends of the secondary of the transformer, points A and E simultaneously. Draw these in Table 2.1. What is the phase difference between the waveforms at the opposite ends of this transformer? This phase difference is _____ degrees.

2.

Observe the voltage waveforms at points A and B simultaneously. Draw these waveforms in Table 2.1, in the same relative positions in the horizontal direction as they appear on the oscilloscope. Measure the peak–to–peak amplitudes of the waveforms and record in Table 2.1. Record the durations of each part of the waveforms also.

3.

There are two ways of explaining the voltage waveform at point B. One is, that when $D1$ is forward-biased by the positive peak of the ac voltage at point A, a circulating electron current flows up through R and thus, by Ohm's law, develops a positive voltage at point B with respect to point F. This voltage would have the same waveform as that of the current flowing through R. The other explanation is based on the concept that a diode acts like an electronic switch. When forward-biased it acts like a closed switch. When reverse-biased it acts like an open switch. Using this concept, when a positive peak is at point A then diode D acts like a closed switch connecting point B to point A. During that part of the sine wave at point A, the oscilloscope, which has a probe connected at point B, is actually displaying the voltage at point A, minus the very small drop across the closed switch ($D1$). During the half–period when a positive peak is at point E, then diode $D2$ would act like a closed switch connecting point B to point E. During the half–periods when negative peaks are at points A or E, diodes $D1$ or $D2$ would act like open switches which would isolate point B from those points during the negative peaks of the ac voltage.

Table 2.1. Full–wave rectifier waveforms without C in circuit.

Point	Two cycles of waveform	Amplitude	Duration of one cycle or pulse
B		V	ms
A		V	ms
E		V	ms

Table 2.2. Voltage waveforms in full-wave rectifiers with C in the circuit.

Point	Two cycles of waveform	Amplitude peak-to-peak	Duration	
B		V	Positive slope	ms
			Negative slope	ms
A		V	One cycle	ms
E		V	One cycle	ms
The dc voltage at B, as measured with a voltmeter was		V.		

Full–Wave Rectifier With C in the Circuit

Turn off the power to the circuit and put in the larger value of C that you used in the half–wave rectifier circuit.

1.

Now with C in the circuit observe and measure the amplitudes and durations of the waveforms at points A and B. In order to see the proper time relationships between the different waveforms they should be displayed simultaneously. Complete Table 2.2. Also record the dc voltage at B.

2.

If you wanted to change the circuit so as to get a negative dc voltage at point B then you would reverse the two diodes and also reverse the polarity of C.

Dual-Polarity Power Supply

Remove C from the circuit and replace R in Fig. 2.1 with two resistances of equal values and use the junction between them as the reference point or chassis ground. See Fig. 2.1.

1.

Without C in the circuit, observe simultaneously and measure the pulses at B and F. Record the results in Table 2.3. Draw two pulses of each in this table.

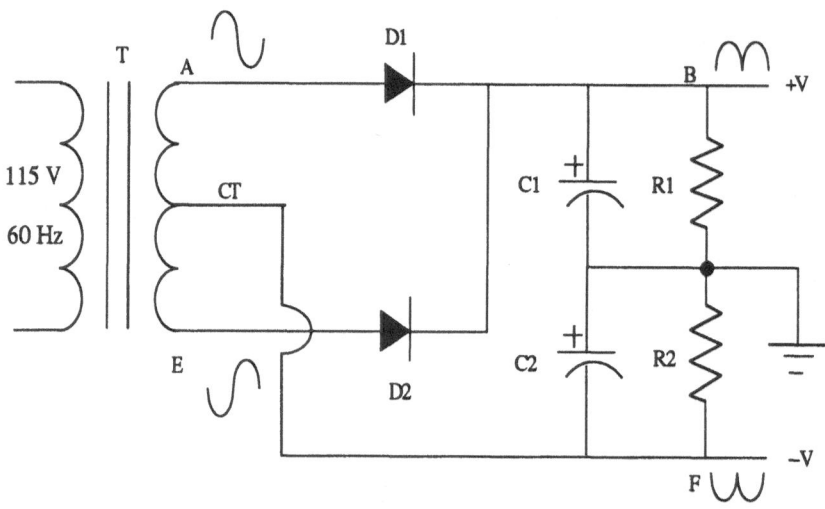

Figure 2.2. Full–wave rectifier with dual-polarity output.

2.

Place the two capacitors in the circuit as shown in Fig. 2.2. Be sure to observe the proper polarities as shown in Fig. 2.2. Then observe the ripple voltages at B and F simultaneously and record the measurements in Table 2.4. Draw two pulses of each in this table. Record the dc voltages at points B and F in Fig. 2.2.

Table 2.3. Voltage waveforms for a dual–polarity full–wave power supply without C in the circuit.

Point	Two pulses of waveform	Amplitudes
B		
F		

Table 2.4. Voltage waveform for a dual–polarity full–wave power supply with C in the circuit.

Point	Two pulses of waveform	Pulse amplitude	Duration when C is being Charged	Duration when C is being Discharged
B				
F				

Questions Concerning the Dual-Polarity Power Supply

1. Were the dc voltages at points B and F equal in magnitude? If they are different, do you think that the difference is significant?
2. What might cause the two polarities of voltage to have different magnitudes?
3. Do you think a single capacitor could have been used instead of the two that we used? If you have time you might want to try it.
4. What did we find about the phase difference between the ac voltage at the two ends of a transformer when an ac voltage is placed across its primary?

3
Full-Wave Bridge Power Supplies

Components

The same as in Experiment 2 but four silicon diodes will be needed.

In this experiment we will build, test, and study a full-wave rectifier that does not require the use of a transformer with a center-tapped secondary in order to function as a full-wave dc power supply. However, in this experiment we will use the center tap in our circuit analysis.

Bridge Rectifiers

Test all four diodes in order to make sure that none of them are defective before building the circuit. This procedure was discussed in the general laboratory instructions at the beginning of this workbook. If this test shows any of them to have a high leakage current, when reverse-biased, then have your instructor replace the defective ones. In this case the ohmmeter would not show a very high resistance when the positive lead of the meter was touched to the cathode of the diode. If you should find a defective diode, ask the instructor to save it so that you can use it at the end of the experiment.

Build the circuit in Fig. 3.1 without capacitor C. After the circuit is built it should be tested before any ac power is connected to the primary of the transformer. A very common mistake made by students is to place one or more diodes in the circuit with the incorrect polarity for the diode. The first step in this test is to clip the negative lead of the ohmmeter to point B in your circuit. Then touch the positive lead of your ohmmeter to point A and then to point E. A very low ohmmeter reading, in the order of 10 Ω or less, should be recorded at each point. This would indicate that $D1$ and $D2$ had been connected properly. A high resistance in either case would indicate that the diode leads should be reversed in the circuit or that the diode was defective. In order to determine if $D3$ and $D4$ are connected properly clip the positive lead of the ohmmeter to point F in the circuit. Then touch the negative lead of the ohmmeter to points A and E. A low resistance reading in each case would indicate that diodes $D3$ and $D4$ were connected in the circuit with the proper polarity.

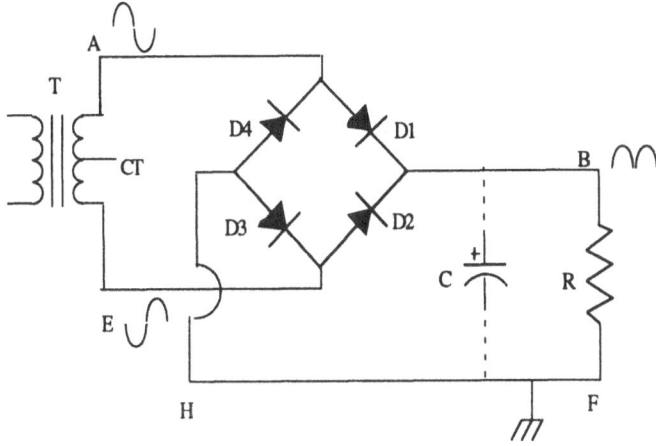

Figure 3.1. A simple full-wave bridge rectifier.

Note that chassis ground in this circuit (point F) is not connected directly to the center tap of the transformer secondary. In fact a center-tapped transformer is not needed for a bridge rectifier to function properly. As mentioned previously, we use the center tap only as an aid in order to study the circuit during this experiment. For this experiment get your external trigger from point A in Fig. 3.1.

1.

Temporarily connect the ground lead of the oscilloscope probe to the center tap (CT) and use it as the reference point. Observe the waveforms at points A and E simultaneously. How do these waveforms compare with those at the same points in the other full-wave power supply you studied? Is the ac voltage applied across the bridge power supply identical to that applied to the two diodes in the previous full-wave power supply? What is the phase difference between the ac voltages at the opposite ends of this transformer secondary?

2.

Now use chassis ground at point F as the reference point. Transfer the ground lead of the probe to that point, and observe the waveforms at points A and E simultaneously. Measure and record in Table 3.1 the items about these waveforms that are listed in that table. Notice the difference in the waveforms at points A and E when the reference point is at point F instead of at the center tap of the transformer secondary. Compare these waveforms to the photographs on p. 34 in the text. For an explanation of this see the discussion of bridge rectifiers on p. 35 in the text.

3.

Next observe and measure the amplitudes and durations of the pulses at point B and record the results in Table 3.1.

Table 3.1. Bridge rectifier waveforms without C in the circuit. Point F is the reference point.

Point	Two cycles of waveform	Amplitude	Duration in ms
B			One pulse
A			One cycle
E			One cycle

1.

Put your larger value of electrolytic capacitor in the circuit. Observe closely the waveforms at points A and E and note any difference between these with C in the circuit and those at the same points before C was added to the circuit.

2.

Now simultaneously display the waveforms at points B and A or E. Measure and record in Table 3.2 the amplitudes of the waveforms. If you notice any anomaly in the waveform at A or E you might try to explain its cause in your report of this experiment. If the sine wave is flattened at its positive peak then you will probably see no anomaly other than its flat top.

  Also measure the duration of the two slopes in the waveform at B, which represent the charge and discharge of C. You might want to improve the accuracy of these measurements by adjusting the sweep rate so that one cycle was expanded horizontally on the oscilloscope screen. Record these measurements in Table 3.2.

3.

Move the sawtooth waveform vertically until it sets on top of the waveform at A or E in order to see the amplitude of the voltage at point A or E when diode $D1$ or $D3$ was forward-biased and when the diode was cut off. The diode would act like a closed switch connecting point B to either point A or point E during the short period of time when $D1$ or $D2$ was forward-biased. The voltage at B would be the same as at point A or E minus the voltage drop across $D1$ or $D2$. This voltage drop across the silicon diode at $D1$ and $D2$ will be measured very precisely in the next part of this experiment.

Table 3.2. Waveform measurements of bridge rectifiers with C in the circuit.

Point	Two cycles of waveform	Amplitude	Duration in ms	
			Charge of C	Discharge of C
B				
			Positive peak	Negative peak
A or E				
Amplitude of dc voltage at point B = V				

Using a Bridge Rectifier to Measure the Voltage Drop Across a Diode During the Time it is in the Conducting State.

1.

Study p. 35 in your text before making these measurements. The simplified drawing in Fig. 2.10 on p. 35 shows the period during which point A is negative and point E is positive. During this period diodes $D4$ and $D2$ are forward-biased. This simplified circuit shows that when the oscilloscope is used to observe the waveform at point A, with point F as the reference, during this period the oscilloscope is actually measuring the voltage across $D4$ while it is conducting current.

2.

Remove the capacitance from the circuit for this measurement. Switch the ac-ground-dc control to ground or short the signal and ground leads together and move the zero-volts line so that it is on the top line of the grid on the oscilloscope screen. Now switch the input to "dc," use point F as the reference, ground lead at that point, and observe the voltage waveform at point A. You will only see the bottom of that waveform. Increase the sensitivity of the vertical amplifier that is being used to 0.1 V/division and measure from the top reference line or zero-volts line down to the bottom of the waveform. Since the resistance R is in series with the diode the actual voltage drop across the diode will be affected a little by the magnitude of R. The voltage drop across this conducting silicon diode = ___V when R = ___Ω.
    You might want to try a number of different diodes or change the value of R.

3.

If a germanium diode is available substitute it for $D4$ and measure the voltage drop across it. If a germanium diode was available, what was the voltage drop across it under these same conditions? V = ___V when R was ___Ω.

Table 3.3. Comparison of different types of rectified power supplies.

Type	dc output	Percent ripple	Relative cost
Half-wave			
Full-wave			
Bridge			

4.

If a defective silicon diode is available you might substitute it for one of your good diodes in a bridge rectifier which includes the capacitor and observe the output voltage at point *B* with the oscilloscope. This might help you recognize this defect when trouble shooting a defective piece of equipment. A high percentage of defects in equipment are in the power supplies because this part of the circuit usually carries more current than the other parts of the equipment.

Summary of Power Supplies

1.

Using the data that you collected in Labs 1, 2, and 3 compare the different types of power supplies. The points to consider include dc amplitude of the output, percent of ripple, which is the ripple amplitude divided by the dc output, and the cost of each type. Use a relative cost such as least or most. Put your data in Table 3.3.

2.

 What do you think would be the result in a radio if a capacitor would become "leaky?" What if the capacitor became "shorted?" How would it affect TV reception?

3.

What do you think the result would be if a diode in the power supply of a radio or TV became "open?" What if it became "shorted?"

4
JFET Amplifiers

Before building, testing, and studying amplifiers, that use junction field effect transistors, in the laboratory we need to have a general discussion of laboratory procedures when working with amplifiers.

Building and Testing Amplifiers in the Laboratory

After building an amplifier connect a waveform generator or sine wave oscillator to the circuit as shown in Fig. 4.1. Most oscillators or waveform generators have an amplitude control incorporated at the output so it is not absolutely necessary that you use a potentiometer like the one in Fig. 4.1. However, in some cases the amplitude that you need to inject at the input of an amplifier must be so low that the signal-to-noise ratio at the output of the oscillator or waveform generator is too low to get a good signal to inject in your amplifier. A voltage divider composed of two discrete resistors could be used instead of the potentiometer. This is useful if you want a precise measurement of the amplitude of the signal being injected at the input of an amplifier.

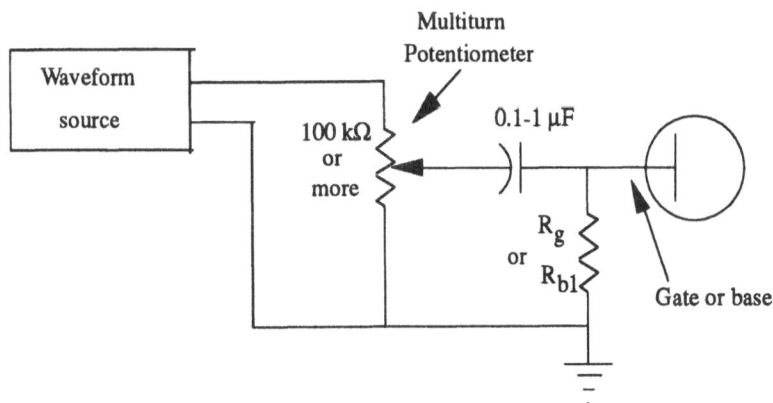

Figure 4.1. Signal source and amplifier.

Testing Your Amplifier

1.

Be sure that the grounds of both the external waveform source and the multiturn potentiometer are connected to the ground of your circuit. Always use a 0.1- to 1.0-µF capacitor between your signal source and your circuit in order to avoid any effect on the junction bias in your transistor.

2.

Get external trigger for your oscilloscope at the square wave output of the waveform generator or the top of the multiturn potentiometer, if using a sine wave oscillator. This will provide trigger voltage with sufficient amplitude to produce stable traces on your screen.

3.

Adjust the amplitude control on your signal source to near maximum and use the multiturn potentiometer to reduce the signal amplitude to the input of your circuit. This should reduce the effect of noise and 60 Hz that might be present along with the signal out of your external voltage source.

4.

Unless the instructions say otherwise, start out with the input of the oscilloscope on "ac." This may make it easier for you to find both traces on the screen.

5.

Connect one probe of your oscilloscope to the side of the coupling capacitor that is connected to the input of the transistor, for example, gate or base. Connect the ground lead of that probe to ground in your circuit. You will not need a ground lead on the other probe.

6.

Always touch the signal probe for the other channel in the oscilloscope to the same spot that the first probe was connected. Both traces on the screen should be identical. This is a test to insure that the oscilloscope controls for both channels are adjusted properly.

7.

Now transfer this second signal probe to the output (drain or collector) of the first stage of your circuit. If this output is inverted and amplified, as compared to the input signal, then follow the instructions for the experiment.

8.

If the output signal is distorted then try reducing the amplitude of the input signal. If the input signal is also distorted try varying the amplitude at the output of the signal source, before it is applied to the top of the multiturn potentiometer.

9.

If there is more than one stage in your circuit always trace the signal from the output of the first stage to the input of the second stage and then to the output of the second stage. Continue tracing the signal from the input to the output of each of the stages in your circuit.

10.

If any stage does not function properly the first thing to do is turn off the dc voltage to the circuit and check to see that the terminals of the transistor (gate, source, and drain or base, emitter, and collector) are connected properly in the circuit. This is a common mistake made by both beginners and experienced technicians. You may need to look in a specifications manual for the diagram of the terminals of the transistor you are using. Remember that this diagram will be the view from the bottom but you will be viewing it from the top. Make sure that your transistor is oriented properly.

Also visually compare your circuit board to the circuit diagram to see that the components are connected in the circuit properly. If this does not solve the problem then proceed in the following manner.

11.

The signal should be the same at the gate or base terminal of the transistor that is at the right-hand side of the coupling capacitor if the two points are connected properly. Use the oscilloscope to see that the signal is reaching the gate or base of the transistor.

IMPORTANT

In "trouble shooting," when told to observe a signal or measure a voltage at a base, emitter, and collector or gate, source, and drain, it means that you are to connect your lead to the actual terminal on the transistor and not to a component that appears to be connected to that terminal. This will often help you discover errors in connecting the transistor in the circuit. This might also help you find a cold solder joint if you are working with a printed circuit board.

There are a few differences between JFET and BJT amplifiers so they will be discussed separately.

Testing JFET Amplifiers

For a JFET amplifier to function as a voltage amplifier there must be a potential difference between the source and drain. This causes a dc electron current to flow from ground up through the source then out through the drain resistor R_d to the power supply. The signal at the gate controls the magnitude of the current through the JFET. Thus a sine wave of voltage at the gate causes the current through the JFET to vary in the same manner. It is the variation of this current through R_d that produces the output signal.

Therefore, for the circuit to function as an amplifier there must be a complete circuit from ground through the JFET and R_d and there must be a potential difference across each component in that path.

Voltage Measurements

The first tests to make on a circuit that does not function properly are voltage tests at the terminals of the transistor. A knowledge of the operation of your circuit would give you an indication of the magnitude and polarity of the dc voltage that should be present at each terminal. If you refer to Fig. 3.5 on p. 49 of your text you can get the approximate dc voltages, with respect to ground, at the drain and source of a JFET amplifier. The voltage at the gate should be zero. Clip the ground lead of your voltmeter to ground or chassis and measure the dc potential, with respect to ground, at the drain, source, and gate of the JFET. If these dc voltages appear normal then use the oscilloscope to see if a signal is reaching the gate. If any of the dc voltages at the terminals are not normal then proceed with the following tests.

1.

With the power on, use a high impedance voltmeter with its ground connected to the ground in your circuit. Measure the dc voltage at the right-hand side of the coupling capacitor. It should be zero volts if the gate resistor R_g is connected in the circuit properly.

2.

Measure the dc voltage at the top of R_d. It should be the same as the power supply. Then measure the dc voltage at the bottom of R_d. If the voltage is the same at both ends of R_d it means that no current is flowing through it. Next measure the dc voltage at the drain terminal of the transistor. If this voltage is not the same as that at the bottom of R_d it means that R_d is not connected to the drain as it should be. If the voltage at the drain is the same as that of the power supply it probably means that either the source is not connected to ground through R_s or that the transistor is defective. If further tests are needed then proceed with the following resistance measurements.

Resistance Measurements

Turn the power off and use the meter as an ohmmeter to measure the resistance between ground in your circuit and the source terminal of the transistor. If your meter is "zeroed properly" the meter reading should be about the same as the resistance of R_s, the source resistor, if this resistor is connected in the circuit properly. If this resistance value is OK then obtain a different transistor from your instructor.

Summary of Voltage Measurements

Now let us go over many of the voltage measurements at the drain, source, and gate and some possible interpretations of those results. These are included in Table 4.1. Table 4.2 lists some resistance measurements and possible interpretations of those results.

Table 4.1. Summary of voltage measurements.

Point of measurement	Measured dc voltage	Possible interpretation
Gate	Zero	OK
	Negative	R_g not connected properly
Top of R_d	Same as V	OK
	Zero	R_d not connected properly
Bottom of R_d	Between zero and V	OK
	Same as top of R_d	No current through R_d
Drain	Same as bottom of R_d	R_d connected to drain OK
	Zero	R_d not connected to drain
	Same as V	Open emitter or defective transistor

Summary of Resistance Measurements

A study of the circuit diagram should give you an idea of the approximate magnitude of the resistance from each terminal of the transistor to ground or chassis in your circuit.

Always calibrate your ohmmeter by shorting the leads together and adjusting the ohmmeter so that it registers zero. Do this before and after making any resistance measurements. It is a good idea to keep on hand a few resistors of known value in order to test the accuracy of your ohmmeter.

Always turn off the power to a circuit before using an ohmmeter to make resistance measurements. When making resistance measurements in a circuit that contains a transistor or diode always make one measurement and then reverse the leads for a second measurement. Use the larger of the two measurements. With one orientation of the leads of your ohmmeter the battery in your ohmmeter may forward-bias a *pn* junction and thus provide a parallel branch to the one that you are trying to measure.

Table 4.2. Summary of resistance measurements.

Point of measurement	Point-to-ground resistance	Possible interpretations
Source	Same as R_s	OK
	Infinite	R_s not connected properly
	Zero	Source shorted to ground
Gate	Same as R_g	OK
	Infinite	R_g not connected properly
	Zero	Gate shorted to ground

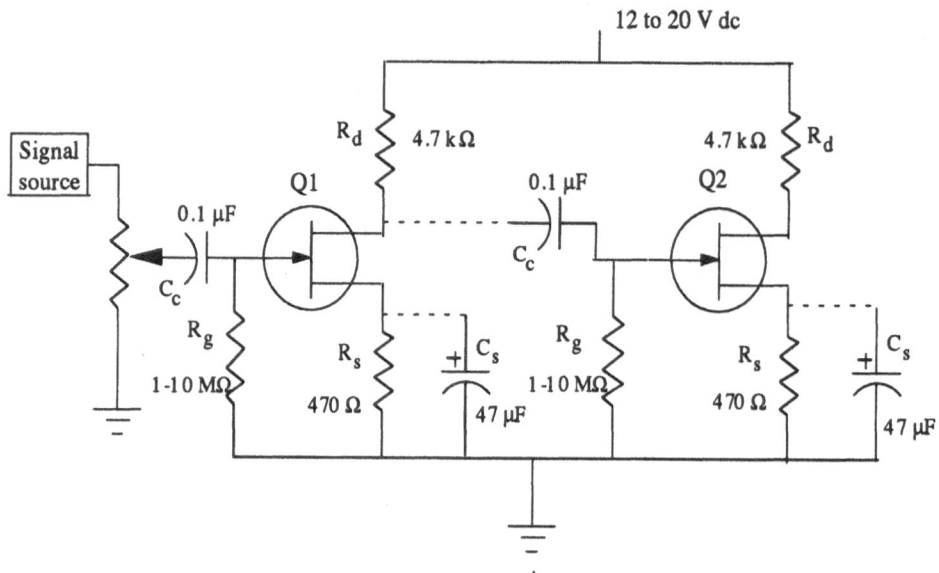

Figure 4.2. A simple two-stage JFET amplifier circuit.

JFET Amplifier Circuit

Components Needed:

1. JFETs: Two N-channel, for example, MPF 102 or MPF 112.
2. Resistors: Two 1 to 10 MΩ (1/4 W), two 470 Ω (1/4 W), and two 4.7 kΩ (1/4 W).
3. Capacitors: Two 0.1 to 0.75 μF with a voltage rating of 50 V and two electrolytic capacitors 47 μF or more with a rating of 50 V.

The magnitudes of these resistances and capacitances are not critical. These are only suggested values.

Before starting each experiment, use a Semiconductor Specifications Manual to find the lead identification diagram of the device that you are going to use. *This is important.* Many times when a circuit does not function properly the cause is found to be a transistor that has been improperly connected in the circuit. The diagram that you find in a manual shows the bottom view of the device but most of the time in the laboratory you install the device from the top. The best way to avoid making a mistake is to draw the diagram as shown in the manual and label the leads with single letters, for example, g, s, and d. Then turn the paper over and place it over a white sheet of paper or window then trace and label the diagram on the back side of the paper to get a top view. Then label the terminals and label the drawing as "top view." You may think that you do not need to go through this simple procedure. However, the author has found that those students that do not use this procedure "sooner or later" build a circuit or circuits that do not operate properly due to a mistake in the identification of the leads of a device.

Single Stage JFET Amplifiers

Circuit Analysis Without C_S in the Circuit

Build only the first stage of Fig. 4.2 at this time. Do not include the source capacitance C_S at this time.

1.

Connect the output of an audio oscillator or function generator to the left-hand side of the coupling capacitor C_c.

2.

Use one channel of your oscilloscope to monitor the signal at the gate and the other channel to monitor the output signal at the drain.

3.

Adjust the function generator, or oscillator, and the oscilloscope so that it displays 5 or more complete cycles of a 1000-Hz sine wave. Get your external trigger at the output of the oscillator. If you are using a function generator use the square wave output as your trigger voltage. Keep the amplitude of the input signal at the gate at 0.1 V (peak-to-peak) or less in order to prevent distorting the output at the drain by overdriving the stage. When you reduce the amplitude of the output of some oscillators to near the minimum, the waveform is distorted and not truly sinusoidal. When this occurs, then connect a potentiometer or voltage divider across the output of the audio generator as shown in Fig. 4.2. In this way it is possible to get a low amplitude sinusoidal waveform without turning the output control of the oscillator to its minimum position. Usually a total resistance of 50 kΩ or more is suitable for the potentiometer.

Note

The author mounted a 10-turn 100-kΩ potentiometer inside a small aluminum box and used 2 BNC connectors for the input and output signals. In this way the amplitude of the input signal could be precisely controlled. The aluminum box served as ground.

4.

Measure the amplitudes of the signal at the gate and drain and record in Table 4.3. Always use peak-to-peak values when measuring ac voltages or pulses unless the instructions tell you to use other values. Calculate the voltage amplification or gain and also record it. Continue to monitor the input signal at the gate while observing the signal at the source. Record the amplitude of the signal at the source in Table 4.3. Compare the amplitudes of the signals at the gate and source without C_S in the circuit. Since the gate is reverse-biased then no current would flow either in or out of the JFET at this terminal. Therefore the same dc current must flow through R_d and R_s in series. When capacitance C_s is not in the circuit then the same ac signal current must also flow through R_d and R_s in series. Therefore the ratio of the signal amplitude at the drain to that at the source should be the same as the ratio of the magnitudes of R_d to R_s. In this case that ratio would be 4.7 k/470 or 10 if you used the suggested values for R_d and R_s. Is the amplitude of the signal at the drain about 10-times that at the source? This would be true, only when there is no source capacitance in the circuit.

5.

What are the phase relationships between the input signal at the gate and the signals at the source and drain? Draw two cycles of the waveform at the gate, source, and drain in Table 4.3 in such a way that these phase relationships are shown.

Note

Plastic templates that have different sine waves are useful for those students that have difficulty drawing sine waves.

Circuit Analysis with C_s in the Circuit

Turn off the dc power to the circuit temporarily while you install C_s in the circuit. Since this is an electrolytic capacitor then the positive lead must be connected at the source of the FET and the negative lead to ground.

1.

Observe the waveforms and measure the amplitudes of the signal at the gate, source, and drain as before. Also record these results in Table 4.3. How does the voltage gain, when C_s was in the circuit, compare with that when C_s was not in the circuit?

Two-Stage JFET Amplifiers

Build stage 2. Connect the second C_c to the gate of $Q2$ but do not connect it to the drain of $Q1$ at this time. Connect your signal source to stage 2 through this coupling capacitor. Make the same measurements on it that you did for stage 1 and record them in Table 4.3.

1.

Transfer your signal source to the input of stage 1 and observe the amplitude of the signal at the gate and drain of $Q1$. It is usually not good procedure to make changes in a circuit while voltage is connected to that circuit.

However, in order to study the effect on the output of the first stage when the input of the second stage is connected to the output of the first we will not follow that rule at this time. While observing the amplitude of the signal at the drain of $Q1$ connect the second C_c to the drain of $Q1$ so that the two stages are connected in series or cascade. Notice whether there was much of a drop in the signal amplitude at the output of stage 1 when the second stage was connected to stage 1. Notice in the circuit diagram that the input impedance of each of these JFET stages would be about 3.3 MΩ. What is the effect of this high input impedance when connecting stages in series?

2.

With C_s in the circuit substitute a "resistance substitution box" for R_d and vary its resistance in order to determine the value of R_d that produces the greatest voltage amplification. Then select a fixed resistance, as near that value as is available, and substitute it for R_d in both stages.

3.

With the two stages in series inject a 1-kHz signal at the input of stage 1. Monitor this input and the output at the drain of $Q2$. Adjust the amplitude at the input of $Q1$ so that the maximum undistorted output is at the drain of $Q2$. Record the amplitudes of the signal at the gate and drain of both stages in Table 4.4. Remove both source capacitances from the circuit and increase the input signal until the output of stage 2 is the maximum undistorted amplitude. Record the amplitudes at both gates and drains in Table 4.4.

Calculate the overall voltage gain of the two stages and compare that gain to the product of the individual gains of the two stages as recorded in Table 4.3.

Table 4.3. Data on two single stage JFET amplifiers with and without the source capacitance.

	Without C_s				With C_s				
	Signal amplitudes				Signal amplitudes				
Stage	Gate	Source	Drain	Gain	Gate	Source	Drain	Gain	
1									
2									
	Waveform					Waveform			
Drain					Drain				
Source					Source				
Gate					Gate				

Were you able to increase the amplification of the stages by changing the value of the resistance of R_d? Compare the maximum amplitude of the input signal, for an undistorted output, without C_s in the circuit with that when C_s was in the circuit.

4.

With both source capacitors in the circuit, use a high impedance voltmeter to measure the dc voltages, with respect to ground, at the gate, source, and drain of each stage. Record these dc voltages on the circuit diagram.

5.

Switch one channel of the oscilloscope to "dc." With ground as the reference observe the dc voltage at the source of stage 1. Then move this horizontal line vertically until it rests on one of the grid lines so that this source voltage can be used as the reference. Now transfer this signal lead to the gate and observe the input signal at the gate. Does the signal at the gate ever become positive with respect to the source when the output signal at the drain is undistorted? Repeat this procedure for stage 2. Remember this result when you perform the experiments on BJTs.

Distortion Due to Large Input Signal

1.

With both source capacitors in the circuit slowly increase the input signal at the gate of $Q1$ until either the top or bottom of the output waveform at the drain of $Q2$ is distorted (flattened). Does $Q2$ go into the cutoff state or the saturated state at this time? Slowly increase the amplitude of the input until $Q2$ is driven into both saturation and cutoff. The output at the drain of $Q2$ should then approach a square wave. What should the voltage be at the drain of $Q2$ during the time that the JFET is in the saturated state? What should the voltage be at the drain during the time that $Q2$ is in the cutoff state? With the input of the oscilloscope on dc, measure the voltage at the drain during the saturated and cutoff states and record below.

Drain voltage while the transistor was saturated $=$____ V.
Drain voltage while the transistor was cut off $=$____ V.

Frequency Response of JFET Amplifiers

You might want to test the frequency response of this two-stage JFET amplifier. In order to do this, monitor the input signal at the gate of $Q1$ and the output at the drain of $Q2$ simultaneously. Inject an input signal with a frequency in the order of 1 kHz and an amplitude that produces an output signal that is not distorted. Choose an amplitude that would be easy to measure and reproduce. Measure the amplitude of the output signal and multiply that amplitude by 0.707 in order to find the amplitude which is down 3 dB. Then find the minimum and maximum frequencies at which the output amplitude does not drop below that 3 dB point when the input is maintained at the original chosen value.

Table 4.4. Signal amplitudes and voltage gains for maximum undistorted output
when the two stages were connected in series.

	Stage 1			Stage 2		
	Signal amplitudes			Signal amplitudes		
	Gate	Drain	Voltage gain	Gate	Drain	Voltage gain
With C_S						
Without C_S						

Laboratory Report

Follow the instructions of your instructor as to the type of report and the material to include in it. You might include in your report the answers to the various questions that were included in these instructions. You might also include a brief discussion of the theory of JFETs and amplifiers that use these semiconductors and some advantages and disadvantages of using JFETs in amplifiers?

5
BJT Amplifiers

Testing BJT Amplifiers

Use the oscilloscope to determine that the input signal is applied to the base terminal of the transistor. If a signal is present at the base but not at the collector then use a high impedance voltmeter to measure the dc voltages in the following order.

1.

The voltage at the top of R_{b2} (the resistor between the base and the power source) should be equal to that of the power supply if R_{b2} is connected to the circuit properly.

2.

The voltage at the junction between R_{b1} and R_{b2} should be positive with respect to ground but only a fraction of that at the top of R_{b2} if the bottom of R_{b1} is connected to ground.

3.

The voltage at the base, with respect to the *emitter*, should be between 0.5 and 0.7 V positive for a signal to be amplified and present at the collector of an npn transistor. You would see no output signal at the collector of a BJT unless the base-emitter bias is forward-biased near the optimum value. If this voltage is not in this range then use a resistor substitution box for R_{b1} (base-to-ground resistor) and adjust its value until the base-emitter voltage is approximately 0.6 V. If, after this adjustment, there is still no output at the collector then make the following measurements.

4.

The voltage at the top of R_c should be that of the power supply if R_c is connected in the circuit properly. The voltage at the bottom of R_c should be positive but much less than that at the top of R_c if current is flowing through it.

5.

The voltage at the collector terminal should be the same as that at the bottom of R_c if that resistor is connected to the collector as it should be.

6.

If the voltage at the collector terminal is equal to that of the power supply and the base-emitter bias is correct then it is probable that the emitter terminal is not connected to ground through R_e or that the transistor is defective. In order to determine which of these is the problem, make the following measurements.

7.

With the power "turned off," measure the resistance between ground and the emitter terminal. If R_e is connected properly this resistance should equal that of R_e, if the ohmmeter has been "zeroed" before making the measurement. If this resistance equals zero, it would indicate that the emitter is "shorted" to ground. If this is the case, and an emitter capacitor C_e is in the circuit, then try removing it from the circuit. If C_e is in the circuit the ohmmeter should register close to zero at the first instant and then rise to the resistance of R_e as C_e is charged up by the battery in the ohmmeter. Capacitors rarely "short out" in low voltage circuits but it does happen once in a while.

8.

If you still have not found the trouble then ask your instructor for a different transistor. The one you have may be defective. In order to test a transistor one should use either a Transistor Curve Tracer or a Transistor Tester. However, an ordinary ohmmeter can be used to get an indication of the condition of a BJT transistor. First you must determine which of the ohmmeter leads is +dc with respect to the other lead. The best way to do this is to use a second dc voltmeter. In order to test a transistor use the R × 100 scale of your ohmmeter. If the transistor is an *npn* then clip the lead of your ohmmeter, that you have determined to be the positive lead, to the base terminal of the transistor. Then just touch, not clip, the negative lead of the ohmmeter, first to the emitter and then to the collector terminals. If the transistor is not defective the resistance reading should be quite low for both the emitter and the collector. Now reverse the leads and repeat this procedure. With the negative lead at the base the resistance at both the emitter and collector should be very high if the transistor is not defective. If the transistor is a power transistor these resistances will be lower than for a general purpose transistor. For *pnp* transistors the leads would be reversed for both trials.

You might want to remove each resistor from the circuit long enough to measure its resistance. With the resistor in the circuit there may be some parallel branches which might make your resistance readings to be in error. It is easy to make a mistake in reading the color coding on some resistors. It is also easy to forget the number associated with each color, or to start counting the color bands at the wrong end of some resistors.

Summary of Testing BJT Amplifiers

Table 5.1. Voltage measurements.

Point of measurement	Measured dc voltage	Possible interpretation
Top of R_{b2}	Zero	R_{b2} not connected properly
	Same as V+	OK
Junction of R_{b1} and R_{b2}	A small fraction of V+	Possibly OK
	A large fraction of V+	R_{b1} not connected to ground
Base terminal	Same as above junction	OK
Base to emitter	About 0.6 V positive	OK
Top end of R_c	Same as V+	OK
	Zero	R_c not connected properly
Bottom of R_c	Fraction of V+	Possibly OK
	Same as top of R_c	No current through R_c
Collector terminal	Same as bottom of R_c	R_c connected properly, ok
	Zero	R_c not connected properly
Emitter terminal	About 0.6 V less than base	OK
	More than base voltage	Emitter not connected to ground through R_e

Table 5.2. Resistance measurements.

Point of measurement	Measured resistance	Possible interpretation
Emitter to ground	Zero	Emitter shorted to ground
	Infinite	R_e not connected properly
	Same as R_e	OK
Base to ground	Same as R_{b1}	OK
	Same as R_e	Reversed ohmmeter leads
	Infinite	Base not connected to ground through R_{b1}

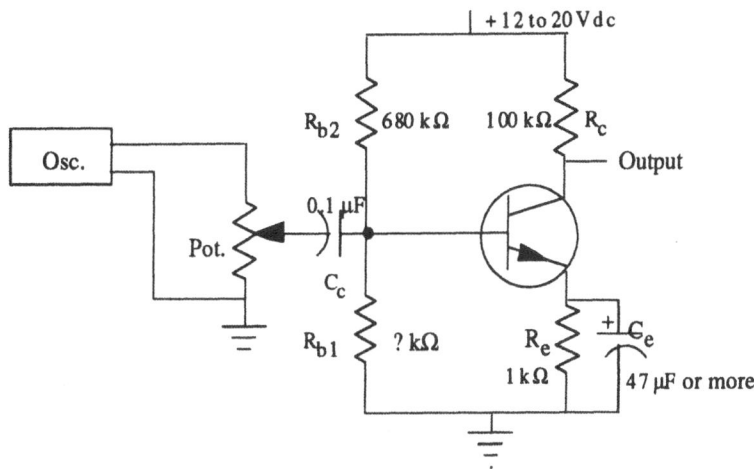

Figure 5.1. Single stage BJT amplifier. V_{be} = _____ V.

Components Needed

1. Transistor: One *npn* bipolar junction transistor, for example, 2N2925, 2N2924, or 2N2712.
2. Resistors: One 680 kΩ, one 100 kΩ, one 47 kΩ, one 1 kΩ, and one to be determined experimentally from a set ranging from 27 to 68 kΩ.
3. Capacitors: One 0.1 µF (nonelectrolytic) and one 47 µF or more (electrolytic).
4. Resistor substitution box. This can be made from a kit if desired.
5. Potentiometer: One 50 to 100 kΩ.

Single Stage BJT Amplifiers

Construction of a Single Stage NPN Common-Emitter Amplifier

Find the data, on the transistor that you are going to use, in a Semiconductor Handbook or Specifications Manual. Note the maximum voltages for this transistor. For your circuit use a power supply voltage of not more than the maximum V_{ce} as given in the manual. Do not let your dc forward bias exceed the maximum value of V_{be}, as given in the manual. Also find the lead identification diagram for the particular transistor that you are going to use. This diagram showing the location of the base, emitter, and collector leads is the bottom view. However, usually when building or working with the circuit you would be looking down on the components so you would need a top view of the lead locations. This is very important. In many years of laboratory supervision by the author, the most common cause of nonoperating circuits was found to be the improper connection of the base, emitter, and/or collector in the circuit. This seems very simplistic. However, usually the student would draw what he thought was the top view while looking at the bottom view in the manual.

After a lot of frustration it would be found that the leads were not properly connected. To avoid this the simple procedure, although given previously, will be repeated.

1. On a sheet of plain white paper draw the diagram of the transistor exactly as shown in the manual and mark the locations of the base, emitter, and collector by B, E, and C. Label this diagram as the bottom view.
2. Turn the paper over and trace the diagram and mark the leads B, E, and C as they are located when looking through the paper. Label this drawing the "top view" and use it in setting up your circuit.

If you have any difficulty in getting any circuit to "operate" as it should then refer to the discussion on testing BJT amplifier circuits or the appendix in the text for a discussion of "Trouble Shooting Electronic Circuits." If at any time, a stage is "overdriven" so that the output is flattened on either the positive or negative peak then reduce the amplitude of the input signal at the function generator or oscillator.

Circuit Construction

Build the circuit in Fig. 5.1 leaving enough room on the circuit board to build two other stages at the right of this one. Include C_e in the circuit. For all of this experiment switch the input of the oscilloscope to ac and get your external trigger voltage from the sine wave output of the external oscillator, or top of the voltage divider if one is used. A voltage divider is sometimes useful at the output of an external oscillator so that an adequate trigger voltage is available even when the signal into the circuit is low in amplitude. This helps to avoid a jitter in the oscilloscope display. If using a function generator then get the trigger voltage from its square wave output.

1.

Measure the dc voltage at the top of R_c and record it at that point in Fig. 5.1.

2.

Use the resistor substitution box, set at zero ohms, as R_{b1}. Connect one probe to the top of R_{b1}, which is also the base of the transistor, and the ground lead of that probe to chassis ground. Connect the output of the sine wave oscillator or function generator to the input of the stage through the coupling capacitor and the ground lead of the sine wave generator to chassis ground of the circuit you have built. Adjust the output of the sine wave generator so that the signal at the base measures about 0.01 V peak-to-peak. Use a frequency of 1 kHz (1000 cycles per second). Use the signal lead of the other probe to monitor the output signal at the collector of the transistor. Use a high impedance voltmeter to monitor the dc voltage, with respect to ground, at the base. Observe the output signal at the collector and the dc voltage at the base as you slowly increase the resistance of R_{b1} (the variable resistance box). Until the resistance of R_{b1} reaches a certain magnitude, the base-emitter junction will be reverse-biased so no output signal will be present at the collector of the transistor.

Table 5.1. Peak-to-peak amplitudes of the signal at the base, emitter, and collector with and without C_e.

	Signal voltage amplitude at			Voltage
	Base	Emitter	Collector	amplification
With C_e				
Without C_e				

3.

If no output signal is present before the dc voltage at the base has reached 1 V then you should start "trouble shooting" the circuit. First, measure the dc voltage between the base and emitter to see if the transistor is forward-biased. Some output signal should be present if the base is positive with respect to the emitter by about 0.55 V or more. Follow the instructions on p. 32 or in the Appendix of the text in "trouble shooting" and testing of components. The author believes that a student often learns more electronics when he has to "trouble shoot" a circuit than when a circuit operates properly when first built.

4.

After your circuit is operating as an amplifier you should use a resistor substitution box as R_{b1} and adjust the resistance of this box to find the optimum value of R_{b1} that produces the maximum amplification of the signal. You may need to reduce the amplitude of the input at the base in order to get an undistorted output sine wave. After you have found the resistance of R_{b1} that produces the maximum amplification then select a discrete resistor as near as possible to that resistance and substitute it for R_{b1}. Sometimes a series or parallel combination of resistors can be used to more nearly approximate the desired value of R_{b1}.

5.

Measure the peak-to-peak amplitudes of the signals at the base, emitter, and collector and record these values in Table 5.1. Remove the emitter capacitor C_e from the circuit and repeat these measurements. Calculate the voltage amplification in each case.

6.

Replace the emitter capacitor in the circuit. Use a high impedance meter to measure the dc voltages at the base, emitter, and collector and record them at the proper locations in Fig. 5.1. Also measure the dc voltage between the base and emitter (forward-bias) and record at the bottom of Fig. 5.1.

7.

Observe the phase relationship between the sine waves at the base, emitter, and collector and draw two cycles of each in Table 5.2 showing these phase relationships.

Table 5.2. Phase of the signals at the base, emitter, and collector in an npn common emitter.

Terminal	Two cycles of sinusoidal signal	
	With C_e	Without C_e
Collector		
Emitter		
Base		

Maximum Input Amplitude for an Undistorted Output

Now let us determine the maximum amplitude of the signal at the base that produces an undistorted output at the collector. Let us compare the result when C_e is in the circuit with that when C_e is removed from the circuit.

Replace the resistance substitution box with a resistor with the optimum value of resistance that you determined experimentally. For the first part of the study C_e should be in the circuit.

1.

Monitor the signal amplitudes at the base and collector with the oscilloscope. Also use a high impedance voltmeter to monitor the dc voltage at the collector during this study. Start with a low amplitude of input signal and slowly increase that amplitude until the output at the collector becomes distorted. Record these amplitudes in Table 5.3. Increase the amplitude of the input signal until the top and/or the bottom of the output sine wave was "flattened" so as to approach a square wave. Did you notice any changes in the dc voltage at the collector when the output signal was greatly distorted? Due to the inertia of the meter movement an ordinary voltmeter would measure the average and not the instantaneous values of collector voltage. Using the oscilloscope, measure the dc voltage on at the collector during the time that the output was "flattened." Explain this phenomenon in your report.

2.

Remove C_e from the circuit and repeat the procedure. What do you notice about the amplitude of the input signal required to cause a distorted output when C_e is not in the circuit compared to the amplitude that caused a distorted output when C_e was in the circuit? When C_e was not in the circuit the signal across R_e served as inverse feedback. How was the voltage amplification affected by this inverse feedback?

Table 5.3. Maximum input for undistorted output in a single stage BJT amplifier.

	With C_e	Without C_e
Maximum output		
Maximum input		
Voltage amplification at maximum input		

BJT Amplifier Frequency Response

In order to determine the frequency response of the amplifier it would be necessary to determine the voltage amplitudes at the output of the amplifier for a wide range of frequencies, when the amplitude of the input signal was kept constant.

A graph would then be made of the amplitudes vs frequencies. The frequency response would be that range of frequencies between those where the amplitude of the output signal dropped to 0.707 times the amplitude in the middle of the range.

Rather than go through this long process let us just use only three frequencies, 50 Hz, 1 kHz, and 10 kHz. An approximate measurement of these frequencies by the oscilloscope would be accurate enough for this study. C_e should be in the circuit. Monitor the amplitude of the signal at the base and collector during this study. Maintain a constant amplitude, for example, 0.01 V, at the base for all three frequencies.

1.

Measure the amplitude of the output at each frequency and record in Table 5.4. Remove C_e and repeat the procedure. You might want to increase the frequency of the input until the output drops off significantly with C_e in the circuit, and C_e out of the circuit, in order to determine the effect of C_e on the frequency response.

Effect of dc Voltage at the Collector on Amplification

If the power supply that you are using has a variable dc voltage output you could determine the effect on the output signal, when the dc voltage at the collector was varied. In doing this, be sure that you do not exceed the maximum voltage on the collector, as listed in a specifications manual. Some graphs of collector current vs collector voltage of typical npn transistors are included in your text starting on p. 402. A study of these graphs indicates that a change in collector voltage has little effect on the collector current.

In these graphs the lines representing the collector current, at each base voltage, are almost horizontal. Therefore a change in dc voltage at the collector should have little effect on the amplitude of the output signal. Does this agree with your experimental result? However, it would have an effect on the maximum amplitude of the input for an undistorted output signal.

Effect of dc Forward Bias on Amplification

Now let us study the effect of the level of forward bias on the amplification of the *npn* common emitter amplifier in Fig. 5.1 with the emitter capacitor in the circuit. Monitor the input signal at the base and the output signal at the collector with the oscilloscope during this study. Use external trigger obtained at the output of the waveform generator, or square wave output if you are using a function generator. Use a high impedance voltmeter to monitor the dc voltage at the base, with respect to the emitter, which would be the forward bias between the base and emitter. Replace the resistance R_{b1} with a resistor substitution box.

1.

Start with the resistor substitution box at its minimum value. Inject a 1-kHz signal with an amplitude of 0.01 V peak-to-peak at the base through C_c. Set the input of the oscilloscope to ac, and the channel that is used to monitor the output at the collector at its lowest volts per division setting, so that you can detect a very small amplitude of signal at the output of the transistor. Start increasing the resistance of R_{b1} slowly until an output signal is observed at the collector. Record the amplitudes of the signal at the base and collector along with the magnitude of the forward bias (voltage between the base and emitter) in Table 5.5. Then slowly increase the resistance of R_{b1} so that the forward bias increases in increments of 0.01V if possible. Enter the data at each increment in Table 5.5 along with the calculated voltage amplification. Maintain a constant amplitude of input signal at the base for all increments of forward bias. Continue increasing the forward bias in increments until the output at the collector becomes distorted. After completing Table 5.5 then graph a curve of amplification vs forward bias (V_{be}).

Table 5.4. Frequency response of a BJT single stage amplifier in the range from 50 Hz to 10 kHz.

Frequency	Input at base		Output at collector	
	With C_e	Without C_e	With C_e	Without C_e
50 Hz				
1 kHz				
10 kHz				

2.

For this study leave C_e in the circuit. Switch one channel of the oscilloscope to "dc" for this entire study. Using ground as the reference observe the dc voltage at the emitter. Then move this horizontal line so that it is on one of the grid lines. This grid line will now be our reference line. Now transfer this signal lead to the base and observe the input signal. Use the other channel to monitor the signal at the collector. Start with the resistance of R_{b1} at the value that a signal appeared at the collector. Gradually increase the resistance of R_{b1} until the output at the collector started to show some distortion. Does the input signal at the base ever go negative with respect to the voltage on the emitter? How does this result compare to that for the JFET amplifier on p. 31?

Table 5.5 represents the amplification of a BJT as a function of the forward bias on its base-emitter junction. Since the number of measurements that you make is unknown then you will have to make this table.

Table 5.5. Amplification as a function of the magnitude of the forward bias on a BJT.

V_{be}	Input at base	Output at collector	Amplification

6
Two-Stage (NPN and PNP) Common-Emitter Amplifier

Components Needed for Stage 2 (in Addition to the NPN Stage in Experiment 5)

1. Transistor: One *pnp* power transistor or any other *pnp* transistor that can be used as an amplifier, for example, 2N3613, 2N4403.
2. Resistors: One 68 kΩ, one 10 kΩ, one 680 Ω, and one to be determined experimentally.
3. Capacitors: One 47 (or more) μF (electrolytic), and one 0.1 μF (nonelectrolytic).
4. Resistor substitution box.

PNP Common-Emitter Amplifier

Find the data on the transistor that you are going to use. Build the second stage in Fig. 6.1, including C_e, at the right of the single stage amplifier used in Experiment 5. Be sure to leave enough room on the circuit board to build another stage between these two stages. This will become the third stage in our three-stage amplifier. Note that the emitter is connected to the positive power supply through the emitter resistor R_e, and the collector resistor or load resistor R_c is connected to chassis ground. This configuration is used so that a single power supply with only a positive voltage could be used for both the *npn* and *pnp* transistors. Be sure to use the proper polarity when connecting C_e in the circuit. The orientation of this stage makes it appear as if this third stage might be an emitter follower but it is a common-emitter stage.

1.

Use a resistor substitution box for R_{b1}, set initially at zero ohms. Using this stage alone, inject a 0.01-V sinusoidal voltage at the input of this stage through C_c. While observing both the input signal at the base and the output signal at the collector simultaneously, increase the value of R_{b1} very slowly until the output signal becomes the maximum amplitude of undistorted sine wave. In this case R_{b1} may have a very low value of resistance, for example, less than 100 Ω when an output signal appears at the collector.

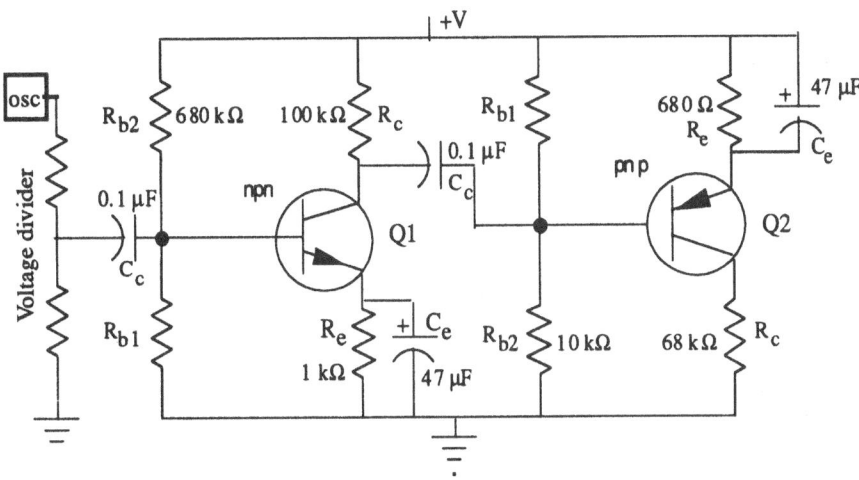

Figure 6.1. Two-stage BJT amplifier composed of *npn* and *pnp* transistors.

Then replace the resistor substitution box with a discrete resistor as near the optimum value as possible. In some cases when substituting a resistance in a circuit, a combination of two resistors in series or parallel can be used to more accurately represent the resistance of the resistor substitution box.

2.

Measure the amplitudes of the signal at the base, emitter, and collector and record these amplitudes in Table 6.1. Remove C_e and repeat these measurements.

Two-Stage Common-Emitter Amplifier

Connect the coupling capacitor of stage 2 to the collector of $Q1$ so that the two stages are in series or cascade. Transfer the output of the waveform generator from the input of the second stage to the input of the first stage. While observing the signal at both the base of $Q1$ and the collector of $Q2$, adjust the amplitude of the input signal so that the output of the second stage is at its maximum undistorted amplitude. If you are using a sine wave oscillator and have trouble triggering the sweep so that the oscilloscope pattern is stable, you might want to use some resistors as a voltage divider in order to obtain the desired amplitude for the input signal, but a greater amplitude of trigger voltage at the top of the voltage divider. This is usually no problem with a function generator because the amplitude of the square wave output is constant. There are two other cases where the use of a voltage divider at the output of a waveform generator is useful. Sometimes the output of the generator is not a good waveform when the amplitude control is near its minimum position. With a voltage divider this control can be at a high amplitude, and still get a low amplitude for use as an input signal. Also, if a very low amplitude of signal voltage is needed, it is often more accurate to measure the voltage at the top of a voltage divider.

Table 6.1. Signal amplitudes and amplification of the *pnp* stage in Fig. 6.1.

	Signal voltage amplitude at			Voltage
	Base	Emitter	Collector	amplification
With C_e				
Without C_e				

Dividing this signal amplitude across the voltage divider by a factor that is determined by the voltage divider, for example, 100 if a 1 to 100 voltage divider is used, is more accurate than measuring the very low amplitude with some oscilloscopes.

1.

Temporarily disconnect the coupling capacitor from the collector of $Q1$. Measure and record the amplitudes of the signal at the base and collector of $Q1$ in Table 6.2.

While observing the amplitudes of the signal at the input and output of stage 1, connect the second coupling capacitor C_c to the collector of $Q1$ so that the two stages are in series or cascade. What happens to the amplitude at the collector of $Q1$ when the input of the second common-emitter stage is connected to it? Explain this effect in your report of the experiment. Was the signal at the base of $Q1$ affected when this connection was made? Now measure the amplitudes of the signal at the bases and collectors of $Q1$ and $Q2$ and record the results in Table 6.2.

Phase Shifts in the Two-Stage Common-Emitter Amplifier

Now let us examine the phase of the signal at the base and collector of each stage. In order to do this you must use the external trigger mode for the oscilloscope. If you are using a function generator get this trigger voltage at the square wave output of the generator. If you are using a signal generator that does not have a square wave output, then get the trigger voltage at the collector of either $Q1$ or $Q2$. The amplitude of the signal at the base of $Q1$ is probably too low to be used as the triggering voltage.

1.

Does each stage have a 180° phase shift from base to collector? Is the total phase shift of the two stages exactly 360°? Is there any phase shift from the collector of $Q1$ to the base of $Q2$? In Table 6.3, draw the waveforms at the bases and collectors of both $Q1$ and $Q2$ in such a way that these phase shifts are shown. Explain these phase shifts in your report.

2.

Compare the amplitudes of the signal at the collector of $Q1$ and the base of $Q2$. These amplitudes are to be included in Table 6.2.

Table 6.2. Effect on the output of stage 1, when a second common-emitter stage was connected to it.

	Signal amplitudes				Amplification	
	Stage 1		Stage 2			
	Input	Output	Input	Output	Stage 1	Stage 2
First stage alone						
With the second stage connected to the first stage						

3.

Now let us measure, in degrees, the phase shift between the collector of $Q1$ and the base of $Q2$. In order to do this expand the waveforms, and move them horizontally, until two positive peaks of the waveform at the collector of $Q1$ are located on vertical lines 10 cm apart. Each cm would then represent $36°$. Estimate the distance, in degrees, from the left-hand positive peak at the collector of $Q1$ to the positive peak on the waveform at the base of $Q2$. If you used one positive peak and one negative peak of the waveform at the collector of $Q1$ then each cm would represent $18°$. Does the voltage at the base of $Q2$ lead or lag behind the voltage at the collector of $Q1$? Explain this in your report. In your report also describe how you determined this phase shift.

4.

Replace the coupling capacitor between the stages with one that has a larger capacitance, for example, 10 μF or larger. Now repeat the study in Step 3 above. Do not draw these waveforms, but explain any differences between these phase shifts and those when C_c was 0.1 μF. Did the use of a larger capacitance effect the loss in signal amplitude between the output of stage 1 and the input of stage 2?

Table 6.3. Phase of the signal at the input and output of each stage of a common-emitter amplifier.

Point	Two cycles of waveform	Phase
Collector of Q2		
Base of Q2		o
Collector of Q1		$0°$
Base of Q1		

7
NPN Emitter Follower

Components Needed

1. Transistor: One *npn*, for example, 2N2925 or 2N2712 or similar amplifier or general purpose transistors.
2. Resistors: One 680 kΩ, one 10 kΩ, one 1 kΩ, and one to be determined experimentally.
3. Capacitors: One 0.1 μF (nonelectrolytic) and one 10 μF or more (electrolytic).

In Experiment 6 it was found that there was quite a loss in signal amplitude when a second common-emitter stage, with its low input impedance, was connected in series with a first stage. One purpose of this experiment is to demonstrate the use of emitter-follower stages to match impedances when coupling common-emitter stages.

Emitter-Follower Circuit

Turn off the power to your previously built two-stage amplifier and disconnect the input of the second stage from the output of the first stage. Build another second stage (emitter follower) in between the two previously built stages, as shown in Fig. 7.1. For R_{b1} use a resistor substitution box, now set at its lowest resistance value. Do not connect this second stage to either of the other two stages yet. Temporarily connect a 10 or more μF electrolytic capacitor from the emitter to ground so as to place the emitter at rf ground while determining the resistance to be used for R_{b1}.

1.

Using a frequency of 1 kHz and an amplitude of 0.01 V inject a sinusoidal signal into the input of this second stage through C_c. Monitor the input at the base and the output at the collector of this second stage. Increase the resistance of R_{b1} until the output at the collector is maximum but not distorted. Then remove the emitter capacitor. The signal present at the emitter should have an amplitude approximately equal to the input signal at the base. The waveforms at the base, emitter, and collector should not be distorted. Replace the resistor substitution box at R_{b1} with a resistor approximately equal to the value found experimentally. Measure the amplitudes of the input signal at the base and the output at the emitter and record in Table 7.1.

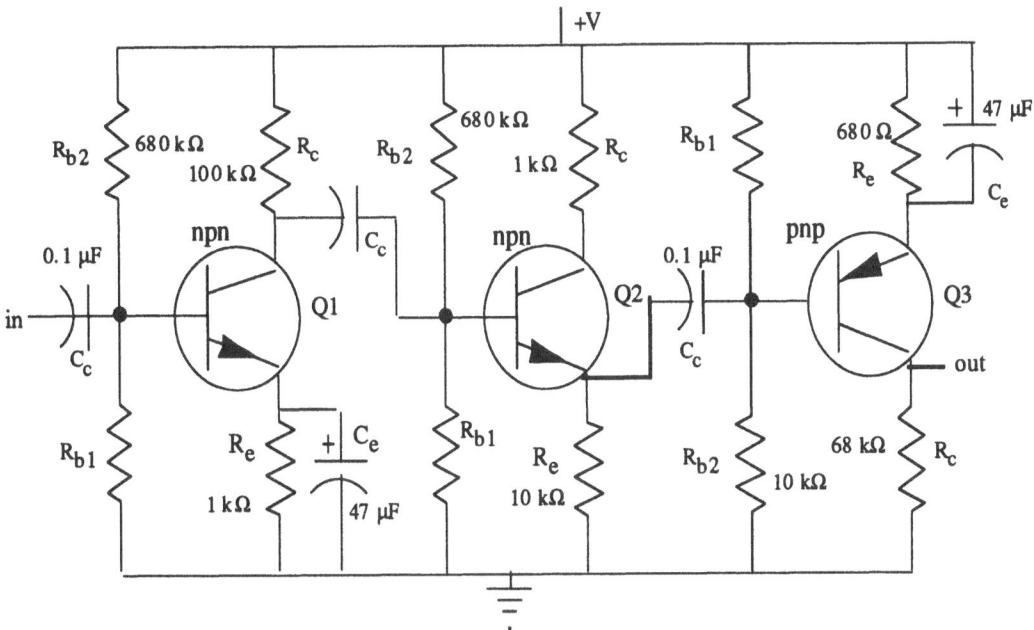

Figure 7.1. Three-stage amplifier using an *npn* common-emitter, an *npn* emitter-follower, and a *pnp* common-emitter in series.

Using the Emitter-Follower Between Two Stages of Common-Emitters

Turn off the power and then transfer the output of the waveform generator to the input of the first stage through C_c. Now connect the input of the second stage (emitter follower) to the output of the first stage and the input of the third stage to the output of the emitter follower. The three stages should now be connected in series or cascade. You should get external trigger at the output of the waveform generator. If using a function generator get it at the square wave output.

1.

While observing the output of the third stage, inject a sinusoidal voltage (1 kHz) at the input of the first stage. Adjust the amplitude of the input signal for the maximum amplitude that would produce an undistorted sine wave at the output of the third stage.

This input signal will probably have a very low amplitude due to the total amplification of the three stages in series. You might want to use a voltage divider made of discrete resistors, for example (1:100), at the output of the waveform generator, as shown in Fig. 4.1 on p. 21. This might help in two ways. If you are not using a function generator you could get the trigger voltage at the top of the voltage divider while you are injecting a fraction of that voltage at the base of $Q1$. You could also measure the amplitude of the signal at the top of the voltage divider and then calculate the amplitude of the signal at the base of $Q1$. When using some oscilloscopes it is very difficult to accurately measure signals at these low amplitudes.

Table 7.1. Signal amplitudes and amplification of an npn emitter-follower.

Circuit	Signal amplitudes		Voltage amplification
	Input	Output	
NPN emitter follower			

2.

Temporarily disconnect the input of the emitter-follower second stage from the output of the first stage. Measure the amplitudes of the signal at the input and output of the first stage and record these amplitudes in Table 7.2.

3.

Monitor the output of the first stage with your oscilloscope. While observing the amplitude of the signal at the collector of $Q1$ connect the input coupling capacitor of the second stage (emitter follower) to the collector of $Q1$. Is there any significant loss in amplitude at the collector of $Q1$ when the second stage was coupled to the first stage.

Three-Stage Amplifier Circuit

With the input of the second stage connected to the output of the first stage all three stages are in series. Measure the amplitudes of the signal at the input and output of each of the three stages and record in Table 7.2. Calculate the voltage gain of each stage and the total amplification of the three stages in series. If the amplifications of the first and third stages are a great deal less than you found in the previous two experiments you might want to use the resistance substitution box and redetermine the optimum value of R_{b1} in each case.

Table 7.2. Input and output signal amplitudes for each stage in a three-stage amplifier consisting of an npn common emitter, an npn emitter follower, and a pnp common-emitter output stage.

Stage in circuit	Signal amplitudes		Voltage amplification
	Input	Output	
Stage 1 before connecting stage 2			
Stage 1 after connecting stage 2			
Stage 2 after connecting to stage 1			
Stage 3			
Three stages in series			
Fraction of the output of Q2 that was injected at the base of Q3			

Phase Shifts Between Stages

1.

Observe and draw in Table 7.3, under the input waveform for the first stage, the output waveforms of all three stages as they appear on the oscilloscope so as to show the phase relationships.

2.

Measure the phase shift between the output of the second stage (at the emitter of $Q2$) and the input of the third stage (at the base of $Q3$). In order to do this observe the signals at these two points simultaneously. The procedure to do this was given in the preceding experiment but will be repeated here. Expand these waveforms horizontally and measure the distance, in millimeters, from one positive peak to the next, in the same waveform. You might want to adjust the small vernier on the horizontal sweep control in order to make the peaks fall on vertical lines on the oscillo-scope screen. This distance represents 360°. Then measure the distance, in millimeters, from a positive peak at the emitter of $Q2$ to the nearest positive peak at the base of $Q3$. Calculate the number of degrees that this distance represents and record in Table 7.3.

Loss in Signal Amplitude When Stages are Connected in Series

1.

Be sure to return the horizontal sweep control to its calibrated position, with the small vernier fully clockwise, if you changed it in the previous measurements. Measure the amplitudes of the output of $Q2$ (at the emitter) and the input to $Q3$ (at the base) and record these values in Table 7.4. Calculate the fraction of the amplitude, of the output of stage 2, that is applied to the input of stage 3 and also record in Table 7.4. This is actually a loss.

2.

Using a high impedance meter measure the dc voltage at the base, emitter, and collector of each transistor and record these voltages at the proper points in Fig. 7.1. Also measure as accurately as possible the forward-bias at the base-emitter junction of each transistor and record below each stage in Fig. 7.1.

3.

The resistances used in the third stage were selected in order to demonstrate the phase shift and loss in amplitude at the input of this stage. The ratio of the capacitive reactance (X_c of C_c) to the resis-tance of R_{b1} at the input of the third stage determines the phase shift and amplitude loss. The use of a larger capacitance and a larger resistance would reduce both the phase shift and amplitude loss. It might be possible to increase the voltage amplification of the third stage by selecting some components with different values. Also this might reduce both the phase shift and loss that was found in the steps above.

Table 7.3. Phase of the signal at the outputs of each stage in Fig. 7.1.

Point in circuit	Waveforms	Phase shift
Input of first stage		0°
Output of first stage		
Output of second stage		
Output of third stage		

Table 7.4. Amplitude loss between stages 2 and 3.

Terminal	Signal amplitude	Fraction of output of Q2 that is injected at the base of Q3
Emitter Q2		
Base Q3		

Possible Contents of Your Report

In your report you might include a comparison of the merits of BJTs and JFETS when used as signal amplifiers and discuss the use of emitter followers in coupling stages of amplification. You might also discuss the phase shifts, and losses in amplitude, when transistor stages are connected in series.

8
JFET Differential Amplifier

Components Needed

1. Transistors: Two N-channel JFETs, for example, MPF 102 or MPF 112.
2. Resistors: Two 1 MΩ, two 4.7 kΩ, and one 470 Ω.
3. Potentiometer: One 2 kΩ.
4. Capacitors: Two 0.1 μF (not electrolytic).

Differential-Amplifier Circuit Using JFETs

·Build the circuit as shown in Fig. 8.1. The values shown are only typical and none are critical, so components that are reasonably close to these values can be used. However, equal values should be used in the two stages. Use a dc voltage source with an output of between +10 and +20 V. Two 9-V transistor batteries, connected in series, could be used.

1.

With no signal at either gate adjust the potentiometer so that the output voltage between points A and B is zero as shown by a dc voltmeter. This is called "balancing the amplifier."

2.

Monitor this output voltage between points A and B with a voltmeter on its most sensitive scale for about 1 min. and note whether there is any output drift. If any exists, then estimate this drift in mV/s. Drift =____mV/s
   Bring a hot soldering iron near (but do not touch) one of the transistors and observe the drift.

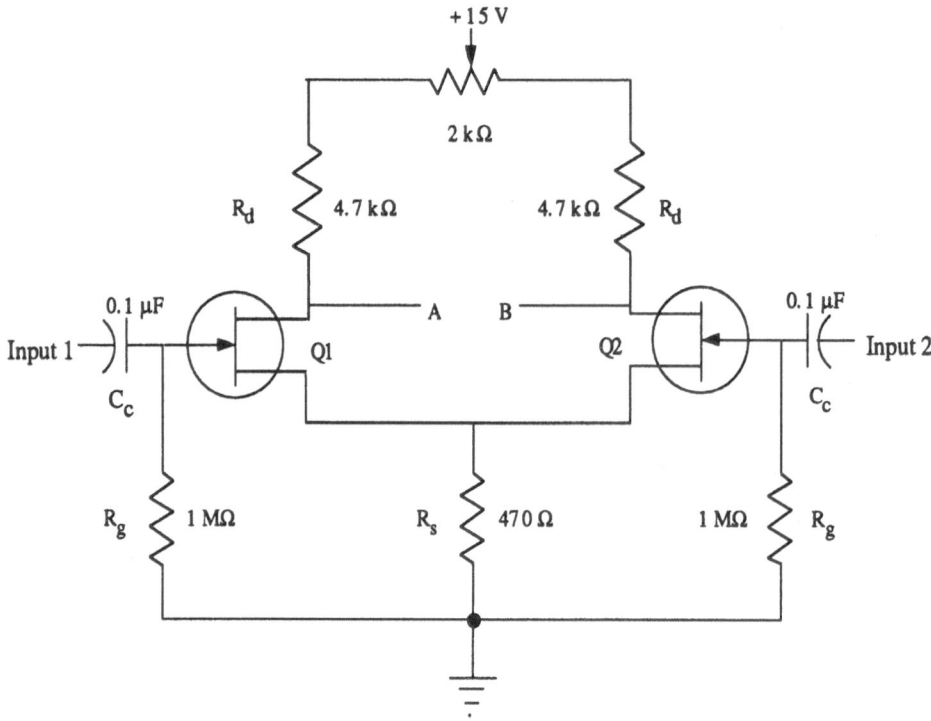

Figure 8.1. A differential or difference amplifier.

Same Input Signal at Both Gates

While observing the output signals at both drains, inject a signal with a frequency of 100 Hz through the coupling capacitors C_c to the inputs of both JFETs. Adjust the amplitude of the input signal so that the maximum undistorted output is present at the drains.

1.

With the oscilloscope measure the amplitudes of the signal at the gate, source, and drain of each JFET and record these amplitudes in Table 8.1.

Difference Amplitude Measurements in Differential Amplifiers

Also measure the amplitude of the output signal between points A and B and record in Table 8.1. You cannot use an ordinary oscilloscope to measure the signal between points A and B directly because you cannot connect a ground lead of the oscilloscope at either point A or B, since neither of these points is at ground potential.

Table 8.1. Results when the same signal was applied to both inputs of a differential amplifier.

JFET	Amplitude of signal		
	Gate	Source	Drain
Q1			
Q2			

The difference signal between points A and B =   V peak-to-peak
The Common-Mode Rejection Ratio (CMRR) =

There are some oscilloscopes that have differential inputs but they are very expensive and thus are not available in most laboratories. However, the following describes three different methods that can be used to determine the output signal between points A and B.

1. Some oscilloscopes have the capability of inverting the signal in one channel and also the capability of adding the signals in the two channels. When a negative quantity is added to a positive quantity, the result is the difference between their absolute values. Therefore, if your oscilloscope does have both of these capabilities then invert the signal at either point A or B and then add this negative signal to the other signal to get the difference between points A and B.
2. You can use an external inverting amplifier (with a gain of 1) to invert the signal at point A or B. Then use an external summing amplifier to add this inverted signal to the other signal and then measure the output of this summing amplifier with the oscilloscope. The circuits of these two special amplifiers were included in the instructions at the beginning of this manual.
3. At this low frequency of 100 Hz you could use a VOM meter to measure the ac voltage between points A and B. However, you must remember that the ac meter registers the voltage in rms volts. Therefore, you would need to multiply this rms voltage by 2.828 (2 × 1.414) in order to convert it to peak-to-peak volts.

Common-Mode Rejection Ratio

The common-mode rejection ratio is a parameter that is used to compare differential amplifiers. It is a measure of the ability of a differential amplifier to produce a useable output signal in the presence of noise or other undesirable signals. It can be calculated from the following equation. It is usually stated in dB. Instructions in calculating dB are included in your textbook.

$$\text{CMRR} = \frac{\text{Input signal common to both gates} \times \text{amplifier gain}}{\text{Output signal between points } A \text{ and } B}.$$

1.

Determine the CMRR for your difference amplifier at 100 Hz and record in Table 8.1.

Phase of the Input and Output Signals

1.

Compare the phase of the signal at the two gates, at the two drains, and between the gates and sources.

Input Signal at Only One Gate

Connect one gate to ground and inject a 100-Hz signal at the other gate through C_c. Select an amplitude that does not produce a distorted output at either drain but large enough to be easily measured at the sources and drains.

1.

Measure and record in Table 8.2 the amplitudes of the signal at both gates, the sources, and both drains (points A and B are at the drains) using ground as the reference. Also measure and record the signal voltage between points A and B in the same way that you did in the previous section.

2.

Show the phase relationships (in Table 8.3) by drawing one cycle of the signal waveform at the different terminals.

Table 8.2. Signal amplitudes in a differential amplifier with one input grounded.

	Q1			Q2		
Parameter	Gate	Source	Drain (A)	Gate	Source	Drain (B)
Signal amplitude						
Amplitude of the signal between points A and B = V_{pp}						

Table 8.3. Phase of the signal at the gates, sources, and drains in a differential amplifier when the signal is injected at one input and the other input is grounded.

Transistor	Gate	Source	Drain
Q1			
Q2			

Optional

A Simple Spectrophotometer

If a light sensitive device such as a cadmium sulfide photoresistor or a silicon voltaic cell is available then the following very simple spectrophotometer as shown in Fig. 8.2 could be built.

If a solar cell is used instead of a photoresistor then the 9-V batteries in the gate circuits would not be needed. However, there must be a complete circuit through the gate resistor in both cases. The polarities of the batteries, or the solar cells, must be such that the polarities of the voltages at the gates are the same. If you do not get much deflection on the meter you might try reversing the polarities of both of the batteries. Place about 40 ml of clear water in each of two small beakers, for example, 50 ml. Then place one beaker on top of the sensitive area of each light sensitive device. Place the complete setup under a fluorescent light so that the light shining on the two light sensitive devices has about the same intensity at both devices.

1.

Use a dc voltmeter on its most sensitive range to measure the voltage between points A and B. Balance the circuit by adjusting the potentiometer in the drain circuits until the voltage between points A and B is zero. Then put food coloring, one drop at a time, in only one of the beakers. The author used green food coloring. Stir the solution and then record the meter reading after each drop. It may be necessary to reverse the leads of the voltmeter between points A and B. Make a table showing the meter reading and the number of drops of dye in the solution.

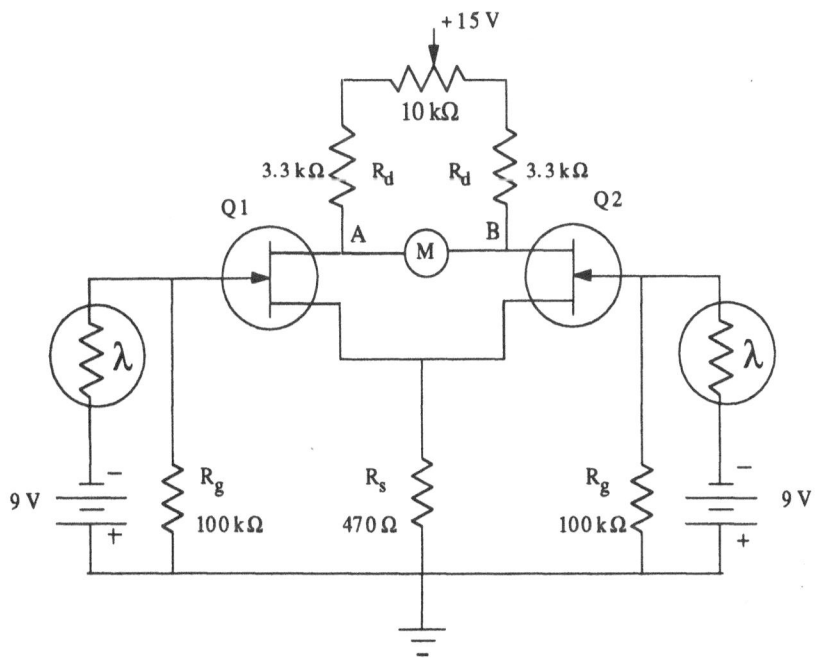

Figure 8.2. A simple spectrophotometer.

Double-Ended Operation of a Differential Amplifier

Figure 8.3 shows a simple circuit that could be used to study double-ended operation of a differential amplifier. This setup could be used to demonstrate how a difference amplifier could be used to discriminate against 60 cycle pickup.

1.

Use a very low frequency signal, for example, 100 Hz, from the oscillator on the right and a higher frequency, for example, 1 kHz, from the left-hand oscillator. If the circuit is balanced correctly there should be some 100-Hz signal at each drain but very little between points A and B because the same low frequency signal is applied to both gates. However, there should be a high amplitude of 1-kHz signal between the A and B because that signal would have a phase difference of 180° at the two gates.

   You might be able to eliminate the right-hand oscillator and use a long, for example, 1 m, piece of unshielded wire connected to the center tap of the transformer. This might pick up enough 60 Hz from the electric field in the room to use for this experiment.

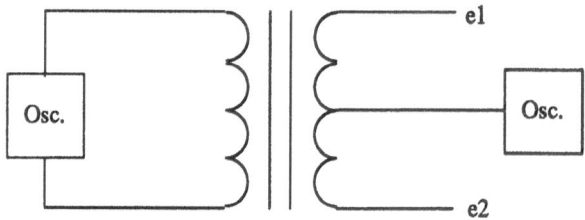

Figure 8.3. A simple circuit that could be used to study double-ended operation of a differential amplifier.

9
Push-Pull Amplifiers

Components Needed

1. Transistors: Three similar *npn* BJTs, for example, 2N2925, 2N2924, or 2N2712.
2. Resistors: One 680 kΩ, two 3.3 kΩ, two 1 kΩ, two 680 Ω, and one to be determined experimentally.
3. Capacitors: Three 0.1 μF (not electrolytic).
4. Potentiometers: Two 100 kΩ.
5. Output transformer: Transistor push-pull type. Two identical ordinary output transformers, without center-tapped primaries, could be used if connected in series. In that case the junction between the two primaries of the transformers would serve as the center tap. The secondaries of the two transformers would be connected in series and serve as one secondary. These secondaries would have to be connected in series in such a way that the output voltages would be in series.
6. Power supply: One that puts out both polarities of dc voltage of equal magnitudes of about +15 and −15 V. Four 9-V transistor batteries, connected in series with the middle junction grounded, could be used. All dc voltages, including the base voltages for Q2 and Q3, can be taken from one power supply.
7. Resistor substitution box.
8. Function generator: Or other sine wave generator.

Push-Pull Circuit

Build the first stage containing $Q1$. Use a resistor substitution box set at its minimum value, instead of the 150-kΩ resistor, for R_{b1}. Inject a sinusoidal waveform of about 10 kHz with an amplitude of about 0.8 V through C_c. Be sure to use external trigger.

1.

Monitor the input signal at the base and one output signal at the collector with the oscilloscope, and the base-emitter voltage or bias with a high impedance voltmeter. Increase the resistance of the resistor substitution box until the output signal at the collector is about equal to the input signal.

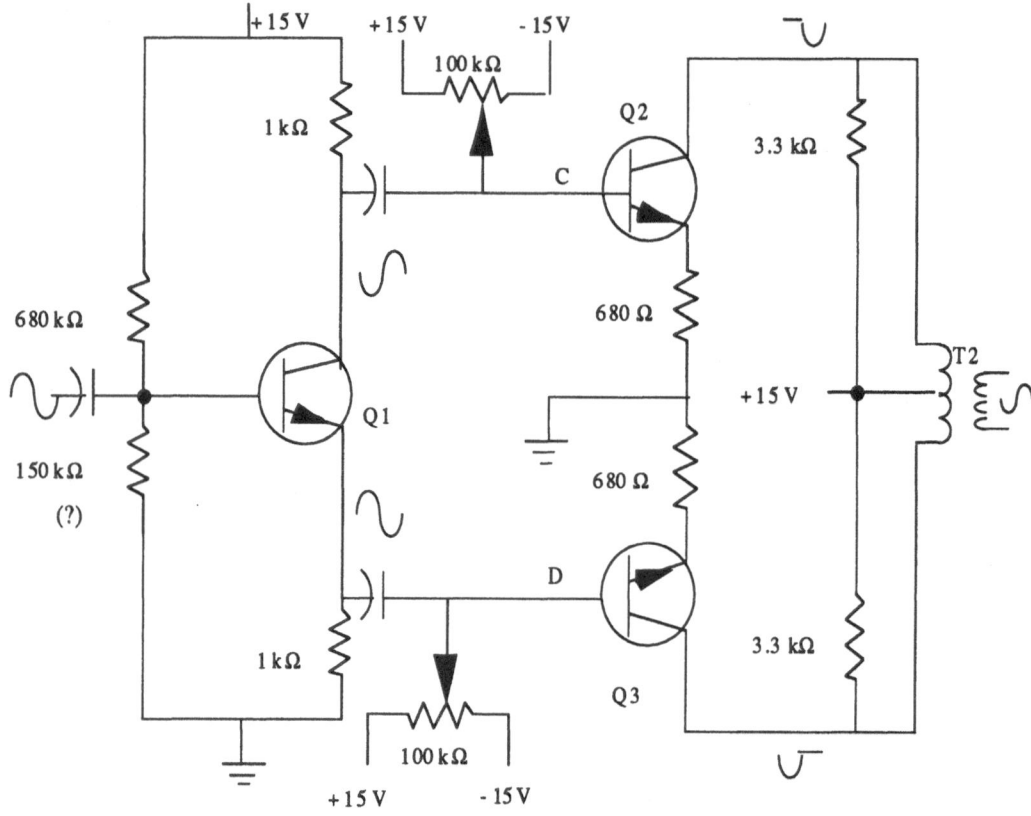

Figure 9.1. Push-pull amplifier with transistor input.

There should be a forward-bias of 0.50 to 0.60 V at this time. If the base-emitter bias exceeds 0.60 V then try reducing it by reducing the resistance of the resistor substitution box. After finding the optimum value for the resistor substitution box then substitute a discrete resistor for it. The signals at the emitter and collector of $Q1$ should be undistorted sine waves.

2.

Note the phase of each of the signals at the base, emitter, and collector. Do they agree with the waveforms shown in Fig. 9.1? Record their amplitudes and the frequency in Table 9.1.

Complete the Circuit

Build the rest of the circuit shown in Fig. 9.1. If there was any high rf signal present along with your injected audio signal it might be necessary to include a 220-pF capacitance between the collectors of $Q2$ and $Q3$. This would be the case if the circuit in Fig. 9.1 was to be connected to the output of the detector in a radio receiver. You might also be able to omit the two 3.3-kΩ resistors in the collector circuits.

The function of these would be to reduce the possibility of saturating the output transformer since each one provides another path for collector current. Their inclusion might make it easier to get the optimum base-emitter voltages in Q2 and Q3.

1.

Use a high impedance dc voltmeter to monitor the voltage between the base and emitter of Q2. Adjust the potentiometer in the base circuit so that there is a forward-bias of about 0.01 V. Repeat this procedure for Q3. This should reduce the distortion that might be present in the signal at the output transformer at the crossover point when Q2 and Q3 change states from conduction to cutoff and vice versa.

2.

Observe the voltage waveforms at the collectors of Q2 and Q3. Each of these waveforms should show one-half cutoff. Compare your waveforms with those in Fig. 5.9 on p. 93 of the text. It might be necessary to vary the base-emitter bias on Q2 and/or Q3 by adjusting the potentiometers in the base circuits in order to get the proper waveforms at the collectors. They should be similar but one should be displaced about 180° from the other.

3.

Observe the audio output signal across the secondary of the output transformer. This should be a reasonably good sine wave. If it does not approach a sine wave in shape then you might try varying the amplitude of the input signal at the base of Q1. If this does not improve the shape of the output then you might try some slight adjustment of the base-emitter voltage(s) of Q2 and/or Q3. There will probably be a small anomaly at the crossover point where transistors Q2 and Q3 change states. You may not be able to remove this small anomaly completely. This is more evident when push-pull transistors are operated as class B amplifiers. In class B the transistors are biased exactly at the cut-off point. This problem is reduced when both transistors are operated with a small forward-bias or as class AB amplifiers.

Table 9.1. Amplitudes of the audio signal at the different points in the push-pull amplifier in Fig. 9.1.

	Q1	Q2	Q3
Base waveform (peak-to-peak)			
Emitter waveform (peak-to-peak)			
Collector waveform (peak-to-peak)			
Positive part of the collector waveform			
Negative part of the collector waveform			
The amplitude of the output signal at the secondary of the output transformer = V			
The frequence of the signal was kHz			

4.

Record in Table 9.1 the amplitudes of the signal at the various points in the circuit. Also measure the dc voltages at the base, emitter, and collector of each transistor and record these values in Fig. 9.1. Compare these dc voltages with those in Fig. 5.8 on p. 92 in your text.

5.

Compare the amplitudes of the positive peaks waveforms at the collectors of Q2 and Q3 with the amplitudes of the negative peaks.

Report

You might want to include some of the following in your report:

1. Why are the signals at the emitter and collector of Q1 equal in amplitude but of opposite polarity? Why is the amplitude of these signals approximately equal to the input signal?
2. Why are the positive peaks of the sine waves at the collectors of Q2 and Q3 cut off or truncated? Why is one of these shifted 180° from the other?
3. How is it possible for the two truncated waveforms at the collectors of Q2 and Q3 to produce a complete sine wave across the secondary of the output transformer?
4. What are the advantages of using the push-pull configuration for an audio amplifier?

10
Transistorized Wien-Bridge Oscillator

Components Needed

1. Transistors: Two JFETs, for example, MPF 102 or MPF 112.
2. Resistors: Two 3.3 MΩ , two 220 kΩ, two 4.7 kΩ, and two 470 Ω.
3. Capacitors: Two 47 µF (electrolytic), two 470 pF (not electrolytic, and one 0.1 µF (not electrolytic).
4. Potentiometer: One 100 kΩ.

The Wien bridge is the series-parallel RC combination at the right of the second transistor in Fig. 10.1. The components in this bridge determine the frequency at which the circuit oscillates. Two stages of amplification are needed in order to get a phase shift of 360° so that the feedback would be positive after point F has been connected to point i. The potentiometer between the two stages is necessary in order to vary the overall voltage amplification of the two-stage amplifier. This overall amplification, from the input to the point at which the feedback voltage is picked off, must be exactly one (unity) in order to produce a sine wave output. If it is less than one then no continuous output would result. If it is greater than one then the output would be distorted and its shape could approach a square wave.

A two-stage amplifier using bipolar junction transistors could be used in front of the Wien bridge instead of the two-stage JFET amplifier in Fig. 10.1. In that case each stage would be similar to the single stage BJT amplifier that you built in the previous experiment. A potentiometer would be used between the two stages in the same way that is shown in Fig. 10.1. Since very little amplification is needed for a Wien-bridge oscillator then the emitter capacitors could be omitted. The author prefers the JFET configuration in Fig. 10.1 because the input of the first stage has a higher impedance and thus has less shunting effect on the Wien bridge.

Theoretical Frequency

Since the two resistors and the two capacitors in the Wien bridge have equal values the theoretical frequency of the oscillator can be calculated by using the following expression:

$$f_0 = \frac{1}{2 \Pi RC} .$$

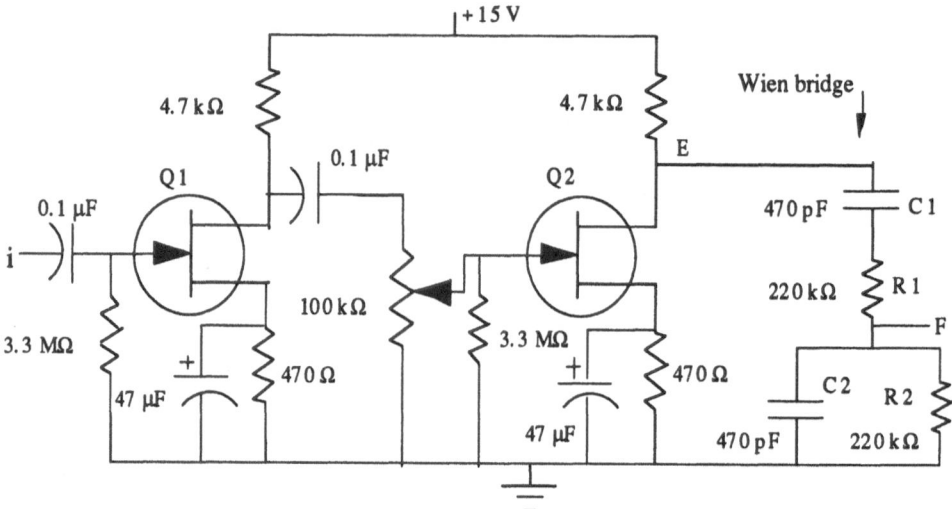

Figure 10.1. Wien-bridge oscillator using JFETs. Positive feedback would be from point F to point i. None of the component values are critical.

Using this expression calculate the frequency of this Wien bridge. Record this calculated frequency in Table 10.1. A small value of C and a large value for R are used in order to avoid the effect of the bottom C, in the Wien bridge, in shunting the signal to ground.

Wien-Bridge Oscillator Circuit

Build the circuit as shown in Fig. 10.1. Do not connect the feedback between points F and i until later. Use an audio oscillator or function generator to inject the calculated frequency at the input of the first stage. Simultaneously monitor this input signal and the output at the top of the Wien bridge, drain of the second stage. Select an amplitude for the input signal that produces an output that is not distorted.

1.

Vary the setting of the potentiometer between the stages and note how it effects the amplitude of the signal at the output. This potentiometer operates as an amplitude control in the same way that a volume control operates in a radio receiver or a TV set.

2.

While observing the input and output signals, vary the frequency of the input sine wave above and below the calculated frequency. Does the amplification appear to be maximum at any particular frequency that is close to the calculated resonant frequency? If this effect occurs, then record this frequency for later reference.

3.

In the Appendix it is shown that the amplitude of the signal at point F should be 1/3 of that at the top of the Wien bridge, point E. Test this assumption by measuring the amplitudes at these two points while the circuit is operating as an amplifier. Use the calculated resonant frequency at the input. Record these measurements and calculations in Table 10.1. Repeat this using a different setting for the potentiometer between the stages.

4.

In order for an oscillator to put out a sine wave the voltage amplification of the circuit must be one. In the Appendix it is shown that the overall voltage gain of the two stages should be exactly three in order to have unity gain at the point at which a positive feedback signal can be picked off. However, when point F is connected to point i the input of the first stage acts as a shunt across the bottom of the Wien bridge. This causes a small loss in the signal so that the overall gain must be slightly more than three in order for the circuit to operate as an oscillator with a sine wave output. In order to test this assumption adjust the potentiometer so that the amplitude at the output of the second stage is exactly three times the amplitude at the input of the first stage. Disconnect the external oscillator from the circuit and then connect point F to point i in order to furnish positive feedback. Does the circuit oscillate with the gain of exactly three? The high input impedance of a JFET makes this shunting effect less than if BJTs were used in our Wien-bridge oscillator circuit.

Circuit Operation or Adjustment

For the rest of your experiment get the external trigger for your oscilloscope at the drain of the second stage.

Table 10.1. Measured and calculated parameters of a Wien-bridge circuit using JFETs.

Conditions		Input to stage 1 at i	Output of stage 2 at E	Voltage gain	Output at F	Fraction V_F / V_E
Using external oscillator	Potentiometer setting 1					
	Potentiometer setting 2					
As an oscillator						
Calculated frequency = kHz						
Measured frequency = kHz						

1.

If your circuit did not oscillate with the gain set at exactly three, it will be necessary to increase the voltage gain by increasing the resistance at the output of the potentiometer between the stages before any voltage waveform appears at the output of the second stage. This adjustment is very critical. If the overall amplification is too great the output will be distorted. If it is too little, any sine wave will be damped or die out. It may be necessary to "fiddle around" with the potentiometer for a considerable time before the output is a constant sine wave that is not distorted. Sometimes it is easier to get the circuit to oscillate when the positive feedback is obtained at point E, the top of the Wien bridge, instead of at point F. Remember, you are trying to get the total amplification from input to the point of feedback to be exactly one, no more and no less.

   The author has never seen a student that did not eventually get a good sine wave from his Wien-bridge oscillator.

2.

Measure the frequency of the output of your Wien-bridge oscillator and record it in Table 10.1.

3.

Measure the amplitude of the sinusoidal output of your Wien-bridge oscillator at the top of the Wien bridge and also at point i and record these in Table 10.1.

4.

Measure the dc voltage at the gate, source, and drain of each stage and record in Fig. 10.1. If your meter does not have a high input impedance then your Wien-bridge oscillator will probably stop oscillating when you try to measure these voltages. In that case delete this step.

   One example related to the Heisenberg Uncertainty Principle is that it is often impossible to measure something without causing a change in the conditions that you are trying to measure.

11
Colpitts Radio Frequency Oscillator

Components Needed

1. Transistor: One *npn* BJT transistor, for example, 2N2925, 2N2924, or 2N2712.
2. Resistors: One 680 kΩ, one 68 kΩ, one 1 kΩ, and one 47 kΩ.
3. Resistor substitution box or linear taper 100 kΩ potentiometer.
4. Capacitors: Two 470 pF, one 0.1 μF (not electrolytic), and two 10 to 100 μF (electrolytic).
5. Copper magnet wire: approximately 22 feet, per lab group, of No. 30 (enamel covered with a varnish-like material).
6. Coil core: One Bic or other slender ball point pen with the tip and ink tube removed to serve as the core of the inductance.
7. One nail or piece of iron or steel wire that would fit inside the ball point pen that is used as the core of the inductance.
8. An rf oscillator: One with variable frequency will be needed for one step in the experiment. If none is available, that step could be omitted.

Colpitts Radio Frequency Oscillator Circuit

It is sometimes difficult to find rf coils with the inductance that one needs. If the desired inductance is not too large, these are often made by winding copper magnet wire on nonmagnetic cores such as plastic or cardboard. Copper magnet wire has a varnish-like coating that serves as an insulation and prevents adjacent windings from forming a short circuit when they touch each other. In order to wind the coil needed for L in the tank circuit in this experiment start with a ball point pen, for example, Bic, and remove the point and ink tube. Some students like to wind a few turns and then wrap masking tape around those turns. Leave about 3 inches of the wire at each end with the rest of the wire wound around the core.

1.

Tightly wind approximately 250 turns of your copper magnet wire around your core, with each winding pushed tightly against the adjacent winding. Make a little loop at the midpoint, after 125 turns, so that you could make a connection at this point if desired.

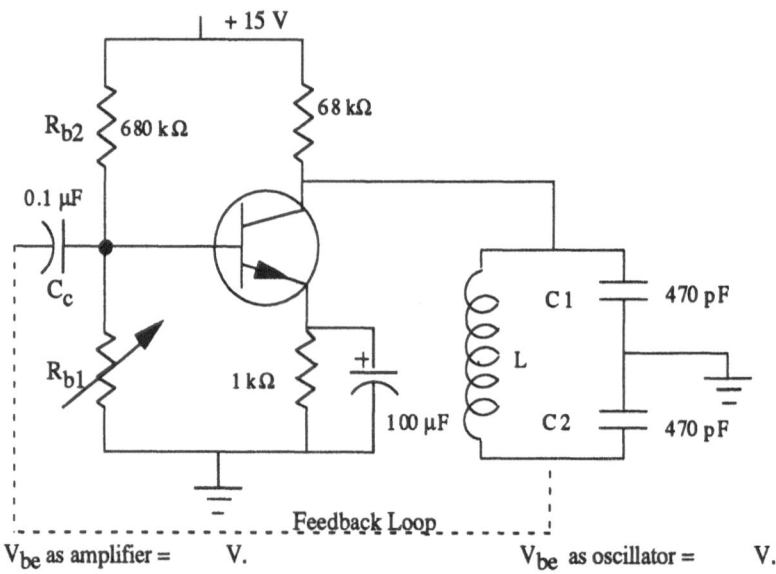

+ 15 V

R_{b2} 680 kΩ

68 kΩ

0.1 µF

C_c

R_{b1}

1 kΩ

100 µF

C 1   470 pF

L

C 2   470 pF

Feedback Loop

V_{be} as amplifier =   V. V_{be} as oscillator =   V.

Figure 11.1. Colpitts radio frequency oscillator.

Wrap a little masking tape, adhesive side down, around each end of the coil to prevent any unwinding. About 3 inches is needed at each end to connect the coil into the circuit. You may need to solder a short piece of larger wire, for example, No. 26 (cloth covered), at each end in order to make good contact in your circuit board. You will need to scrape off the enamel on the magnet wire where it is to make contact with your circuit or if you want to solder it to a piece of the larger wire. The diameter of a Bic ball point pen barrel is such that 250 turns of No. 30 magnet wire will provide an inductance of approximately 55 µH. The mathematical expression used to calculate the inductance can be found in many Electrical Engineering Handbooks. See p. 389 in the text. The inductance of any coil is proportional to N^2 where N is the number of turns. If your coil has more or less than 250 turns you can use this relationship to calculate the inductance L of your coil.

2.

Calculate the resonant frequency (f_0) of your parallel resonant circuit by using the expression

$$f_0 = \frac{1}{2 \, \Pi \sqrt{LC}}.$$

(11.1)

In this expression, for Fig. 11. 1, C is the capacitance of *C1* and *C2* in series. The calculation of the equivalent capacitance of two capacitors in series is similar to that of the resistance of two resistors in parallel, that is, product/sum. When the two components are equal in value, as in Fig. 11.1, the resulting value is one-half of either so two 470-pF capacitors in series have an equivalent capacitance of 235 pF.

Circuit Construction

Build the circuit in Fig. 11.1 with one modification. We are going to study the circuit as an amplifier before we add the positive feedback in order to make it an oscillator. Therefore, it will be necessary to change the ground connection, that is shown at the junction of $C1$ and $C2$, to the bottom of the LC resonant tank circuit. You will need to use an electrolytic capacitor, for example, 47 μF or more between the bottom of the tank circuit and ground in order to avoid shorting the dc voltage, at the collector, to ground. Use an external rf oscillator to inject the calculated rf at the input of the amplifier through the coupling capacitor.

1.

If no rf generator is available then disconnect the tank circuit from the collector and use a generator tuned to its maximum frequency. Monitor the input at the base and the output at the collector. Use a resistor substitution box, or a potentiometer as a variable resistance, for R_{b1} and adjust it for maximum amplification. Leave it in the circuit if you wish, or replace it with a resistor that has the same value of resistance.

2.

With the tank circuit connected to the collector, as described under circuit construction above, use the oscilloscope to observe the output at the collector and the input at the coupling capacitor. Get your trigger voltage at the collector. Start with the external oscillator tuned to a frequency that is about one-fourth of the calculated resonant frequency.

   While observing the amplitude of the rf signal across the LC resonant circuit, at the collector, slowly increase the frequency of the external oscillator. The amplitude of the output signal across the resonant circuit should increase and then decrease as you pass certain fractions of the resonant frequency. A very sharp maximum should occur when you pass the resonant frequency of the LC circuit. Continue increasing the output frequency of the external oscillator until you have a frequency that is more than three times the calculated resonant frequency of the LC circuit.

   You should observe more maxima, rise and fall, in the output signal across the LC circuit. These maxima will vary in amplitudes. The greatest maximum amplitude should occur at the resonant frequency of the LC circuit. Note and record the "reading" on the external oscillator dial at this greatest maximum. Compare this "reading" with the frequency as measured on the oscilloscope. These frequencies should be the same at this one maximum.

   Unless you are using an external oscillator that has been calibrated recently, it is probable that the frequency measurements made with your oscilloscope are more accurate than those shown on the oscillator dial. Record in Table 11.1 the "readings on the dial" of the external oscillator, as well as the frequencies as measured on the oscilloscope, that all of these maxima occur and draw a circle around the frequency that the greatest maximum occurs. At this frequency the external oscillator frequency and the frequency as measured on the oscilloscope should be equal. This is the measured resonant frequency of the LC tank circuit and should be reasonably close to the frequency that your Colpitts oscillator puts out when you connect the feedback loop.

   Did you find that the frequency of the signal across the LC tank circuit, as measured by the oscilloscope, was the same at all of the maxima ? This would indicate that the frequency across a resonant circuit is determined by its circuit components, at all of the maxima, and not by the frequency of the external oscillator. This also indicates that it is possible to excite a resonant circuit with frequencies that are fractions and/or multiples of its resonant frequency. For this reason it is possible to use resonant circuits as doublers or triplers of frequencies. This is useful when the desired output frequency is so high that, if a crystal is used to control it, the crystal would be so thin that it would be too fragile to use. A thicker crystal could be used to control a lower frequency and *LC* resonant circuits used to double or triple that frequency.

Base-Emitter Bias for a Colpitts Oscillator

The base-emitter dc voltage on a transistor when used in an oscillator circuit is usually different than that when used in an ordinary voltage amplifier. This voltage is determined by the *RC* time for discharge of the feedback capacitor and the period between positive peaks in the feedback wave-form. It often has a negative polarity but it may sometimes be positive.

1.

Use a high impedance voltmeter to measure the dc bias at the base using the emitter as reference. As an amplifier the base-emitter junction should have a dc forward-bias of approximately 0.59 V. Record your measurement in Fig. 11.1.

This circuit has only one transistor, which provides a 180° phase shift. In order to get an in-phase signal for positive feedback there must be a method of providing another 180° phase shift in the ac signal. We found in our study of power supplies that the two ends of a coil have a phase difference of 180° when the midpoint of the coil is used as the reference. The same phase difference can be produced across an *LC* tank circuit by using either the midpoint of the inductance, or the capacitance, as the reference point. If we connect one end of the tank circuit to the collector of the transistor and the other end to the input or base of the transistor we would get the total 360° phase

Table 11.1. Data on the Colpitts oscillator when the circuit was excited by an external oscillator.

Parameter		Frequency (MHz) and/or amplitude at maxima						
Frequency and/or voltage		Below resonance			At resonance	Above resonance		
		1	2	3	4	5	6	7
External oscillator dial reading								
Frequency as measured	At input							
by the oscilloscope	At collector							
Amplitude	At input							
	At collector							
Voltage amplification								

shift necessary for positive feedback in order to produce oscillations. The Colpitts oscillator in Fig. 11.1 places the midpoint of the capacitance at ground potential. A Hartley oscillator uses a similar circuit but places the midpoint of the coil at ground potential. In that case a capacitance must be used between the midpoint of the coil and ground to avoid shorting out the dc voltage at the collector.

Colpitts Oscillator Circuit Operation

Change your circuit into an rf oscillator by removing the ground that you used at the bottom of the tank circuit, grounding the junction between the two 470-pF capacitors and connecting the bottom of the tank circuit to the input coupling or feedback capacitor. This is the feedback loop in Fig. 11.1.

In a Colpitts rf oscillator, current flows through the transistor during only a fraction of the period of the waveform. During the positive peak of the feedback waveform the coupling capacitor is charged up rapidly by base current through the forward-biased base-emitter junction. During the rest of the period of the feedback sine wave the coupling capacitor discharges through the parallel combination of R_{b1} and R_{b2}. The RC time for discharge would be much longer than for the charge through the base-emitter junction. This would cause the base-emitter dc voltage to be less positive than when the circuit was operating as an amplifier. In many cases this dc voltage on the base will be negative with respect to the emitter.

1.

Measure the dc voltage at the base, with respect to the emitter, with a high impedance voltmeter and record this bias for your oscillator in Fig. 11.1. If it stops oscillating then disconnect and reconnect the power supply. That should shock the circuit back into oscillation.

2.

Simultaneously observe and measure the frequency and amplitude of the rf oscillations at the base, and at the collector, and record in Table 11.2. Get your external trigger at the collector. Is the amplification unity, as it should be for an oscillator to produce sine waves? Are they in phase or 180° out of phase?

3.

Switch the input of one channel of the oscilloscope to dc and then observe the dc voltage at the emitter. Move this straight line, that represents the emitter voltage, until it rests on one of the grid lines of the screen. Then observe the feedback signal at the base with that probe. Using this emitter volts line as the dc reference measure the period and the amplitude of the positive peak of the feedback voltage at the base. This is the time during which the base-emitter junction is forward-biased. What is the state of the transistor during this positive peak? Repeat this for the negative peak. Record these measurements in Table 11.2. What is the state of the transistor during the negative part of the signal? If the transistor was cut off during this negative peak why is there a signal at the collector during that time?

Table 11.2. Data on the circuit when used as a Colpitts oscillator.

	Frequency	Amplitude
At the collector		
At the base		
Maximum forward-bias		
Total period that the transistor was forward-biased		
Maximum reverse-bias		
Total period that the transistor was reverse-biased		
Calculated value of the inductance of your coil of wire = µH		
Assumed value of the inductance of your coil of wire from p. 66 = µH		

4.

Calculate the inductance of your coil by substituting your measured frequency of oscillation in Eq. (11.1) p. 66 and record in Table 11.2.

Effect on Base-Emitter Bias When the Amplitude of the Output of the Colpitts Oscillator Changes

During this next part of the experiment, monitor the base-emitter dc voltage or bias with a high impedance meter and the rf signal at the base and collector with the oscilloscope. Record in Table 11.3 this dc bias voltage V_{be} and the frequency of the signal at the collector.

1.

Insert a nail or piece of iron wire as far into the core of the coil as possible and repeat these measurements and record in Table 11.3. It is important to make these measurements as precise as possible. In your report explain the differences in these two parameters caused by insertion of the iron into the core. While observing the signal, slowly move the iron slug in and out of the core of the coil and note its effect on the frequency of the oscillator. Repeat this while observing the effect on the base-emitter bias on the transistor.

Most intermediate frequency (i.f.) transformers in radio and television receivers have sintered iron slugs in their cores that can be moved in or out in order to vary the resonant frequency of the primary and/or the secondary of those transformers.

Stray Capacitance in a Coil of Wire

In this last part of this experiment let us show that there is a capacitance between the turns of the coil of wire that we used as the inductance in this experiment.

Table 11.3. Effects on an oscillator due to an insertion of an iron slug in the core of the inductance.

	Without iron core	With iron core
Frequency		
dc reverse bias		

Table 11.4. Data on the oscillator circuit using the coil of wire as both the capacitance and inductance.

	Frequency	Amplitude
LC circuit using coil only		
Calculated value of the capacitance between the windings of the coil, using the calculated value of L on p. 70		

1.

To do this, remove the two 470-pF capacitors, and the ground wire at their junction, from the circuit. The circuit should oscillate, but at a higher frequency than when the two capacitors were in the circuit. Measure the frequency and amplitude at the collector and record in Table 11.4. In this case the coil of wire furnished both the inductance and capacitance for the *LC* resonant circuit.

2.

What happens if you ground the midpoint of the coil of wire with an electrolytic capacitor that has a large capacitance, for example, 100 µF ?

3.

Using the calculated value of inductance from Table 11.2, and the frequency of oscillation when the resonant circuit was composed of the coil only, use Eq. 11.1 to calculate the capacitance associated with your coil made of 250 turns of wire. Record in Table 11.4.

Report

You might want to include the answers to the following questions in your report:

1. What determined the frequency of our Colpitts rf oscillator?
2. What would happen to the reverse-bias on the oscillator if the amplitude of the output changed?
3. What do you think caused the oscillations to start?
4. What controls the amplitude of the oscillations?
5. What is the effect of inserting of an iron core in a coil?

12
Measurement of Unknown Frequencies

Components Needed

1. A source of frequency to be measured. You could use the Colpitts oscillator that you built in experiment 11.
2. A variable frequency rf generator.
3. An audio amplifier.

An abbreviated form of this experiment could be completed using only your Colpitts oscillator, or your waveform generator, and your oscilloscope.

There are a number of methods that can be used to measure the frequency of a signal. The method used often depends on the equipment that is available. In this experiment the students can use as many of the following methods that he desires, if the necessary equipment is available. You can use the output of the Colpitts oscillator, that you built in the previous experiment, as the unknown frequency to be measured.

Using the Oscilloscope to Measure the Period of the Unknown Frequency

The most common method is to measure the period (T) of the signal, with an oscilloscope, and then calculate the frequency using the following expression:

$$f = \frac{1}{T}.$$

Be sure that the small vernier control on the sweep frequency control is at its maximum clockwise position. In this position the sweep is calibrated and the dial setting of the sweep control is correct.

1.

Inject the unknown frequency in the input of the oscilloscope. Adjust the sweep frequency of the oscilloscope so that only one complete cycle of the waveform is on the oscilloscope screen and measure its duration (period). If the unknown waveform is symmetrical you could display only one-half of the waveform, for example, the positive half, and measure its duration (a half-period). Record your results in Table 12.1. All of the tables are at the end of this experiment.

Frequency Meters

1.

If a frequency meter is available the frequency can be read directly from it. Follow the directions for the use of that particular meter and measure the same unknown frequency as before and record the results in Table 12.2.

Lissajous Figures

The horizontal sweep controls on most oscilloscopes have one setting that is labeled "Horizontal Input," x-y, or some other label that has the same meaning. Some have an x-y setting on one of the other controls. When at this setting, one of the channels in the oscilloscope is connected to the horizontal amplifier. Under these conditions the sweep rate is determined by the frequency and linearity of the signal applied to that channel and the internal sweep oscillator is turned off.

1.

Switch both channels to the ac input mode. Use a variable frequency oscillator along with the source of the frequency to be measured. Use the variable frequency generator to inject the horizontal signal, and the unknown to inject the vertical signal. If the external variable frequency oscillator is adjusted so that its frequency is the same as the unknown frequency, and if the V/cm controls are adjusted properly, the resulting figure should be a circle, if there is a phase difference of $90°$. The result would be a straight line if the two signals were equal in frequency and in phase. If the frequency of one of the signals is twice that of the other the resulting figure would be a "figure 8."

Adjust the frequency of the variable frequency oscillator until the pattern on the oscilloscope is stable and does not roll. Adjust the V/cm controls of both channels so that the width and height of the Lissajous figure are approximately equal.

Count the number of times that the Lissajous figure crosses the x axis and the number of times it crosses the y axis. The ratio of these "crossings" is the ratio of the two frequencies applied to the two channels of the oscilloscope.

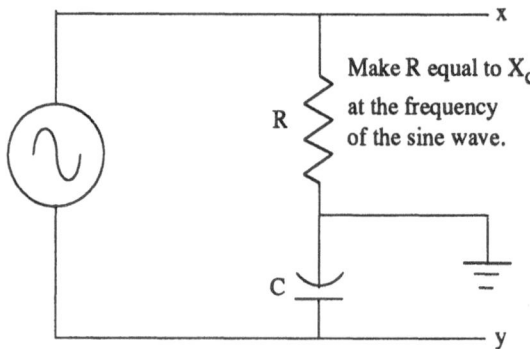

Figure 12.1. Circuit used to produce a circle on the oscilloscope.

It is necessary that one of these frequencies be known in order to calculate the frequency of the other by using this ratio. Record the known frequency, from its dial setting, and the x and y "crossings" in Table 12.3.

Intensity Modulation

Many oscilloscopes have a receptacle on the back side that is often labeled "Z axis," "Z input," etc. When a sinusoidal voltage is injected at this receptacle, one of the peaks of this signal will enhance the brightness of the trace on the screen of the oscilloscope and the other peak will dim the trace during that peak. If the intensity control on the front of the oscilloscope is set at a low level and the amplitude of the input at the Z axis is great enough, a black dot would appear in the waveform during that peak of the input to the Z axis. Thus there would be one black dot in the trace for each complete sine wave of the signal at the Z input.

1.

Build the circuit in Fig. 12.1. It can be used to produce a constant circle on the oscilloscope.

2.

Use the x-y mode of input to the oscilloscope for this experiment. Use a variable frequency oscillator to inject an ac signal across the circuit in Fig. 12.1. Adjust the V/cm controls of each channel so that the figure on the oscilloscope screen is a circle. Use a low intensity for the trace. Inject the frequency, to be measured, at the z axis receptacle. Slowly vary the frequency of the signal across Fig. 12.1. If the amplitude of the unknown signal at the z input is high enough, black dots will appear in the circle. When the unknown frequency of the signal, at the z axis, is an exact multiple of the signal producing the circle these dots in the circle will be stationary. Count the number of dots in the circle. Multiply the frequency of the oscillator across Fig. 12.1 by the number of black dots in the circle to find the frequency of the signal at the z input and record in Table 12.4.

Zero Beat of the Signal, with the Unknown Frequency, with a Signal that has a Known Frequency

When two frequencies are mixed, or added across a nonlinear component or device, the resulting signal has four components. These four components include the two original frequencies along with their sum and their difference frequencies. When the frequencies of the two original signals are close enough together, the difference component is in the audio range. This audio component can be amplified and then detected by the human ear. Audio amplifiers are nonlinear devices. Therefore, if we use an audio amplifier to mix the two frequencies we could hear the difference frequency when the two frequencies are close enough for that difference frequency to be in the audio range. As the frequencies of the two signals approach each other the pitch of the difference signal becomes lower and lower and sometimes resembles the beat of a drum. When the two signals have the same frequency the difference would be zero, the beats would stop, and no sound would be heard.

Some universities have an old U.S. Navy surplus item (BC 221) that was designed for the measurement of the frequency of unknown signals. It included a crystal that could be used to calibrate its output frequency. Thus, when zero beat was achieved the frequency of both signals, being the same, could be determined quite precisely.

1.

For this study any variable frequency rf oscillator, such as the BC 221, could be used. Inject both a signal from the output of this oscillator and the signal, whose frequency is to be determined, at the input of an audio amplifier. A loudspeaker or headphones should be connected to the output of the audio amplifier. Vary the frequency of one oscillator, with the known frequency, until a high pitch sound is heard at the output of the audio amplifier. Then vary this frequency in such a way that the pitch of the sound from the audio amplifier becomes lower and lower. Continue to vary this frequency until you hear a few beats per second. Then slowly continue until these beats disappear. This is the point of zero beat and the frequencies of the two signals are exactly equal. All you need to do is to read the dial of the variable frequency oscillator in order to determine the frequency of the unknown signal. Record your results in Table 12.5.

If an audio amplifier is not available a radio receiver could be used. An automobile radio, that has been removed from a car, is a good device to use. All voltages are low and it is usually easy to reach the volume control. Tune the radio so that no station output is audible. Connect the two signals, to be beat, across the volume control of the radio. Adjust the volume control so that the difference signal is audible. Then proceed as with the audio amplifier in the preceding paragraph.

Table 12.1. Frequency calculation from the period measured by an oscilloscope.

Period of unknown frequency	Calculated frequency of unknown

Table 12.2. Frequency measured by a frequency meter.

Frequency of the unknown signal as measured by a frequency meter was

Table 12.3. Using Lissajous figures to measure an unknown frequency.

	Crossings of x axis	Crossings of y axis	Known frequency	Measured frequency
Parameter				

Table 12.4. Using intensity modulation to measure an unknown frequency.

	Frequency at Z input	Number of dots	Known frequency	Measured frequency
Parameter				

Table 12.5. Frequency as measured by the zero-beat method.

Frequency as measured by zero beating with a known frequency =   Hz

13
Printed Circuit Boards

Materials, etc., Needed

1. One light-sensitive copper coated (on one side) printed circuit board about 3.5 inches long and 2.5 inches wide.
2. No. 2 photoflood lamp.
3. Chemicals: Trichloroethylene, ferric chloride, and acetone (most finger nail paint removers contain acetone).
4. Drill bit: size number 68 to 64 (diameter 0.031 to 0.036 inch).
5. A mechanical stirrer is helpful but not necessary.
6. A dark well-ventilated area during part of the process.
7. The list of materials needed for the circuit to be mounted on the printed circuit board can be found at the beginning of the next experiment.

Printed circuit boards are coated with a thin layer of copper on one side. In a few cases both sides are copper coated. This copper layer is covered with a light-sensitive emulsion. In the process of making a printed circuit, most of the copper must be removed so that the remaining copper forms a circuit that serves to connect the components so that the desired electronic circuit would be made. A photographic negative which contains the pattern of the desired printed circuit is placed over the light sensitive board and a bright light is directed through the negative onto the board. The pattern of the printed circuit is transparent on the negative while the rest of the negative is opaque. After exposure, the light-sensitive emulsion under the opaque parts of the negative must be dissolved and removed from the board before the whole board can be exposed to light. The copper that is no longer covered by the emulsion can then be dissolved and removed from the board. The only copper that remains is that which forms the printed circuit.

There are a number of methods that can be used in order to make the negative that contains the diagram of the printed circuit. Kits are available that contain all of the materials needed to make printed circuits.

Making a Printed Circuit Board

The method we will use includes the following steps:

1. Draw the schematic of the electronic circuit that is to be put on a printed circuit board. This is shown in Fig. 13.1.
2. Draw an outline of a printed circuit that would connect the components in the same way as in the actual circuit. For large circuits this is often quite difficult. This diagram, for an astable multivibrator, is included in Figs. 13.2, 13.3, or 13.4.
3. Make a negative that includes the diagram. It should be the same size as the printed circuit that is needed.

Using Our Printed Circuit Board

Fig. 13.1 is the actual circuit of the astable multivibrator to be built. This figure is at the end of these instructions for this experiment.

1.

Photograph the pattern in Fig. 13.3, at the end of these instructions, and use the negative to make the printed circuit for the above astable multivibrator. If you have a camera that uses roll film, that is large enough, or cut film then take a full size picture of the diagram in Fig. 13.3. Have it developed and use the negative in the first step of the following procedure.

There are other more direct methods of making a negative for use in the exposure of a printed circuit board. If a copier is available, that has the capability of making transparencies, then it can be used to make the negative from either Fig. 13.3 or Fig. 13.4 which has the white diagram on the black background. The circuit pattern must be transparent and the rest of the negative must be opaque. Use this as the negative in the exposure of the printed circuit board. It is also possible to make this negative with a computer system if it includes a scanner and a laser printer.

Procedure

1.

In the procedure that follows, perform the first two steps in total darkness, or while using a photographic safe light. These first two steps should also be performed near a hood or in a room with an exhaust fan since the organic chemical to be used in step 2 is somewhat toxic.

1. Lay a photographic negative of the desired pattern on the light sensitive board and place a piece of flat glass over it. Expose for 6 min using a No. 2 photoflood lamp at a distance of 16 inches from the board.
2. Wash in trichloroethylene for 2 min. while agitating the liquid. This stops the action by removing the rest of the light sensitive material that has not been activated by the light. Avoid inhaling the fumes by using a hood or exhaust fan. This chemical, like most organic solvents, is toxic.

The rest of the procedures can be performed under daylight conditions. It is not necessary to use a hood for the rest of the development of the printed circuit board.

3. Pick it up by the edges and keep it upright until dry.

4. Let it set overnight.
5. Etch with ferric chloride for, from 20 min to an hour, in order to remove the bare copper that is not covered by the exposed emulsion. It should be agitated frequently during this time. Magnetic agitators that use rotating magnetic fields are available. The use of these speeds up the process especially in this step. Other agitators, including small air pumps that use air bubbles to agitate the solutions, are also available.
6. Wash in warm water in order to remove the ferric chloride from the board.
7. Wash the copper parts (printed circuit) with acetone until you see the residue leave and the copper pattern looks bright. If you do not get all of the residue off the board then solder will not adhere to it.
8. Use a drill bit size 68 to 64 (diameter 0.031 to 0.036 inch) to drill a hole in the center of each circular dot in the copper diagram.

Mounting Circuits on Printed Circuit Board

We will now use the completed printed circuit board by placing an astable multivibrator circuit on it. The list of components is included at the beginning of Experiment 14. The circuit diagram (Fig. 13.1) is at the end of these instructions.

1.

Place the components on the side of the printed circuit board that does not have the copper diagram. Figure 13.2 shows the location of the components on the board. Be sure that the transistors are oriented properly.

Now solder the leads of the components to the printed circuit. Do this on the side opposite the mounted components. Use the smallest amount of solder that is necessary to connect the wires to the copper diagram. Be sure that the solder does not "short" between the terminals of the transistors. After soldering then cut off the excess wire. Be sure that the wire does not move before the solder solidifies. You can speed this up by blowing your breath across the hot solder joint. If the wires move during the time that the solder is "setting up," a cold solder joint may result. These are quite often the causes of malfunctioning circuits.

Astable Multivibrator Circuit on the Printed Circuit Board

Use a 9-V transistor battery snap-on connector to connect the power to the circuit. Solder the positive lead of this 9-V transistor battery snap-on connecter to the point marked +9V in Fig. 13.2 and the negative lead at the point marked −V. Be sure to have the proper polarity, Solder short leads at the output terminals in order to connect to oscilloscopes and/or other circuits.

1.

With the battery connected to the circuit, connect an oscilloscope to the output of the multivibrator at the collector of one of the transistors of the circuit. If the outputs at the two collectors are good square waves then proceed to experiment 14. If the circuit does not put out a square wave at the two outputs then do the following. After each step check to see if the circuit is functioning properly.

1. Disconnect and then reconnect the 9-V battery. Sometimes this will "shock" the circuit and cause oscillations. Do this any time that the circuit quits oscillating.
2. Check to see that the transistors are connected properly in the circuit.
3. Use a high impedance voltmeter to see that positive dc voltage is present at the collector of each transistor.
4. With the battery disconnected from the circuit, use an ohmmeter and the actual circuit diagram to see that the components are connected properly in the circuit.
5. With the battery connected to the circuit, inject a signal (it can be sinusoidal) at the base of one transistor while observing the output at the collector of that transistor with an oscilloscope. If the transistor is not defective it should amplify and invert the signal that is on the base.
6. If none of the above solves the problem then touch a hot soldering iron to each solder joint in the circuit. Do this long enough to melt the solder at each connection. Be careful not to let the wire move until the solder is solid. This should get rid of any cold solder joint that might be present.
7. If the dc voltages and resistances measurements are normal but the circuit is not oscillating after you have disconnected and reconnected the 9-V battery then replace one or both of the transistors.

If any component needs to be removed from the circuit hold a solder wick against the solder joint and then heat the joint with a soldering iron. The wick will absorb most of the melted solder and make it easier to remove the component. Then use long nose pliers to gently pull on the component on the opposite side while heating the solder joint.

Other Techniques in Making Printed Circuit Boards

There are printed circuit kits available that employ other techniques. When using one of these then follow the instructions that are included with the kit being used.

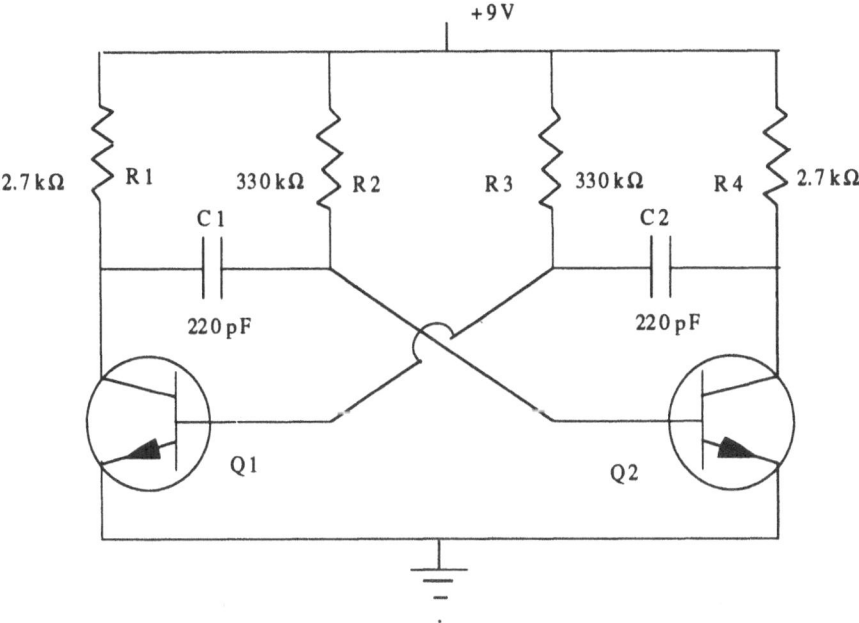

Figure 13.1. Astable multivibrator circuit.

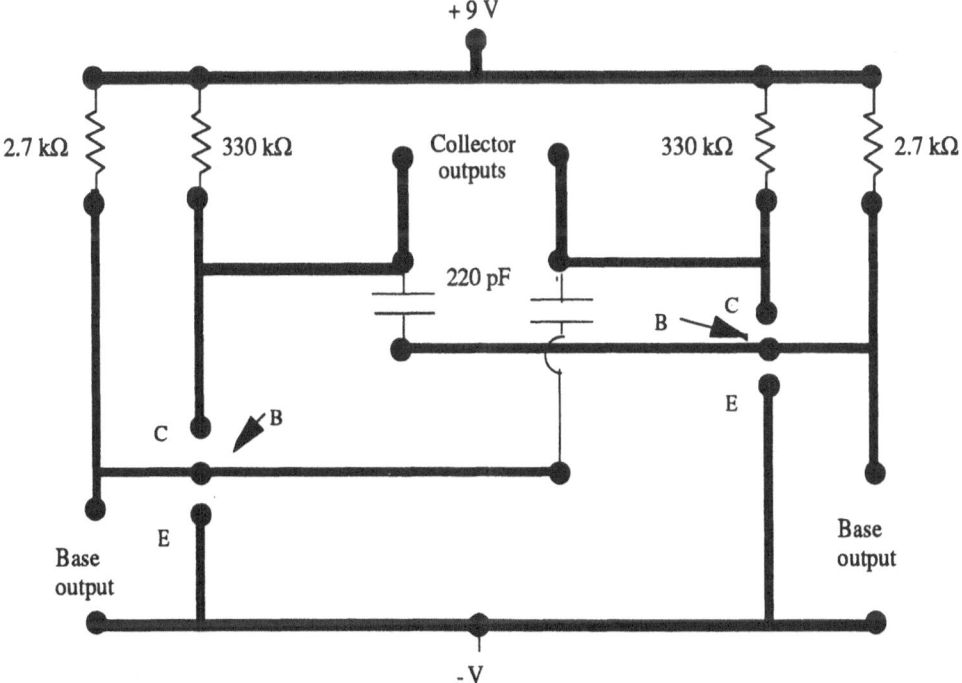

Figure 13.2. Top view (opposite side from the printed circuit) of the printed circuit board showing the location of the components.

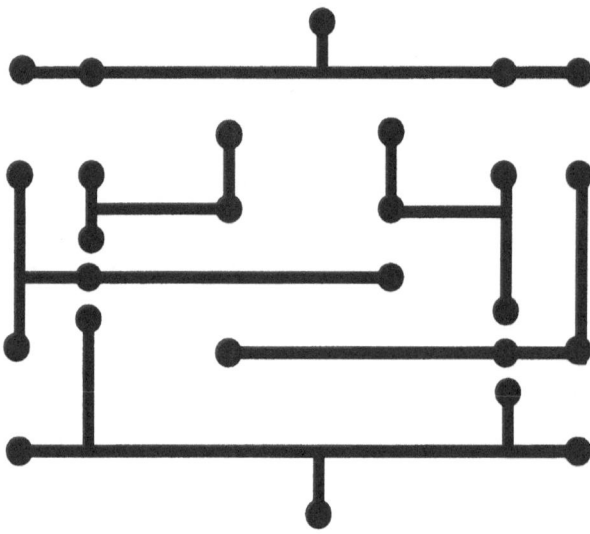

Figure 13.3. Pattern for the negative to be used to expose the light-sensitive copper board.

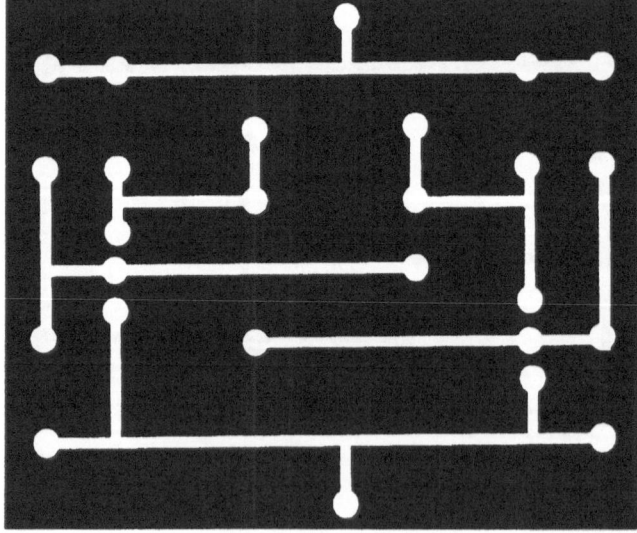

Figure 13.4. Alternate pattern that might be used to make the negative needed to make the printed circuit board.

14
Astable Multivibrator

Components Needed

1. Transistors: Two *npn* BJTs, for example, 2N2925, MPS2222A or any general purpose *npn* transistor.
2. Resistors: Two 2.7 kΩ, two 330 kΩ (these values are not critical but each pair should be two with the same values).
3. Capacitors: Two 220 pF (these values are not critical but should be two with the same values).
4. One 9-V transistor battery and one clip-on connecter for this battery.

Astable Multivibrator Circuit

If you built the astable multivibrator on the printed circuit board in the previous experiment then you can use it for this experiment. If you did not do this, then build the circuit in Fig. 14.1 on a solderless breadboard.

1.

If you are building a new circuit for this experiment then do not connect *C2* to the collector of *Q2* until later. Without *C2* connected to the collector of *Q2* the circuit will be a simple two-stage amplifier with *C2* serving as a coupling capacitor to the base of *Q1*. You can use a 9-V transistor battery as the power source. Inject a low amplitude sinusoidal signal through *C2* to the base of *Q1*. Get your external trigger for the oscilloscope at the base of *Q1*. Observe the signal at the base of *Q1* with one channel of the oscilloscope. Use the other channel to trace the signal from the base of *Q1* to the collector of *Q1*, the base of *Q2* and the collector of *Q2* in that order. Each transistor should invert and amplify the signal.

If either one does not amplify and invert the signal use a voltmeter to see that positive dc voltage is present at the collector. Also check to see that the transistors are connected in the circuit properly. If the proper signal is present at the collector of *Q2* then disconnect the external oscillator and connect the feedback capacitor *C2* to the collector of *Q2*. Use the oscilloscope to see if square waves are present at the collectors of *Q1* and *Q2*. If at any time the circuit quits oscillating then disconnect and then reconnect the dc power supply to the circuit.

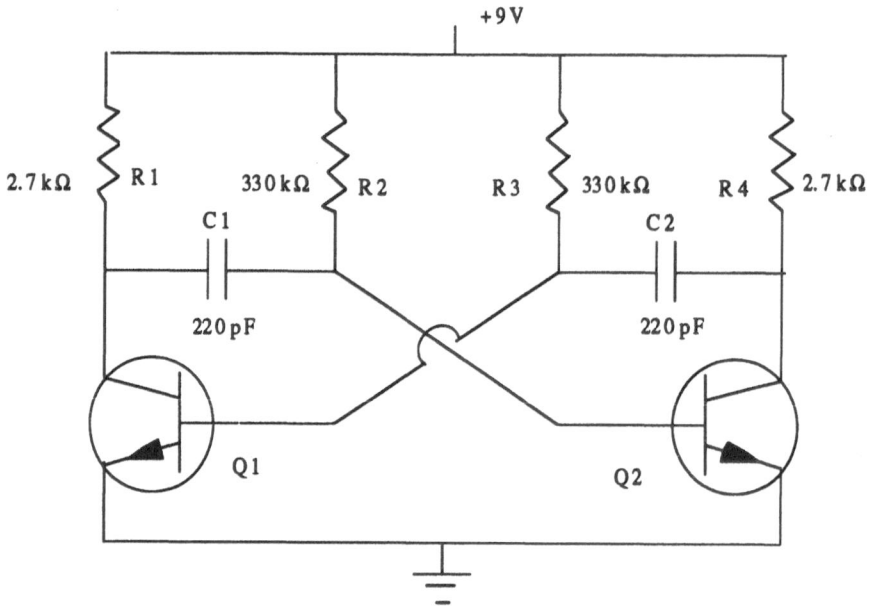

Figure 14.1. Astable multivibrator circuit.

This will usually cause the circuit to start oscillating if the circuit had been oscillating and then stopped for no apparent reason.

2.

Now get your external trigger voltage at the collector of either *Q1* or *Q2*. Switch the input of both channels in the oscilloscope to ground or "short" the signal probes to ground in order to locate the zero volts reference lines. Move these horizontal straight lines vertically until one sets on the second grid line (counting from the bottom) and the other one on the grid line just above the middle line. After you have located the zero volts reference line for each channel then switch the inputs for both channels to "dc." Then observe the waveforms at the two collectors simultaneously.

One cycle of square waves similar to the waveforms at *Q1* and *Q2* are included in Table 14.1. If any corners of your waveforms are "rounded" then make those changes in the square waves at the top of Table 14.1.

3.

Measure and record the dc voltage (with the oscilloscope on dc) during the positive and negative peaks of the waveform at each collector. Record these voltages at the ends of the dashed lines extending from the square waves in Table 14.1. Are the transistors in the cut-off state during the positive peaks, and in the saturated states during the negative peaks? If a transistor is in the saturated state the voltage drop across the transistor would be approximately 0.3 V. If it is in the cut-off state the voltage at the collector would be the same as the voltage source.

Table 14.1. Data on the square waves at the collectors of the transistors in an astable multivibrator.

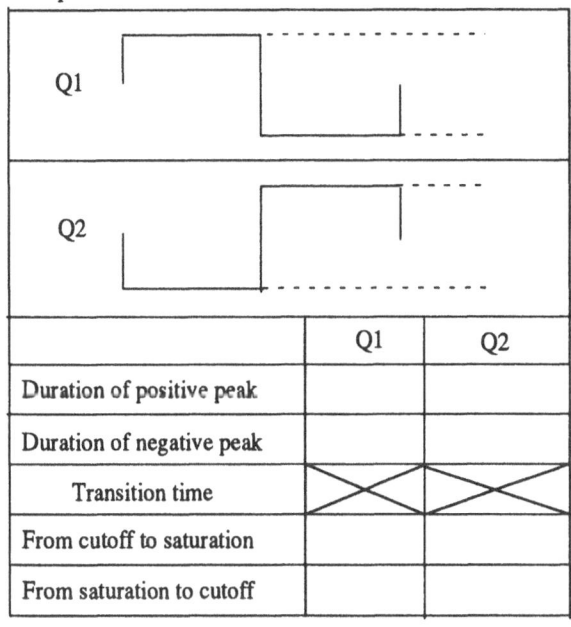

	Q1	Q2
Duration of positive peak		
Duration of negative peak		
Transition time		
From cutoff to saturation		
From saturation to cutoff		

4.

Measure the duration of the positive and negative peaks at each collector. Record this, and all of the other data collected in this section, in Table 14.1.

5.

Are all of the corners of the waveforms square? If any are "rounded" can you explain the reason?

6.

Measure and record the time required to make each transition (from cutoff to saturation and vice versa) for either transistor. This portion of the curve appears to be vertical unless the horizontal time base (time per division) is expanded. If this line is curved at either end then measure the time from the 0.1 maximum point to the 0.9 maximum point. This is the accepted procedure in those cases, especially if the curve gradually approaches a minimum or a maximum.

Waveform Analysis

For this part of the study switch the input of the oscilloscope to ground and move the dc zero volts line from the second grid line to the middle grid line but leave the other dc zero volts line at the grid line just above this middle line. Get your external trigger voltage at the collector of $Q1$. Switch both of the inputs of the oscilloscope to dc. On the top half of the oscilloscope screen display the voltage waveform at the collector of $Q1$.

On the bottom half of the screen display the voltage waveform at the base of *Q1*. The voltage waveform at the base is a typical exponential curve resulting from the discharge of a capacitor. In this case it is the discharge of *C2*.

1.

Measure the duration of the exponential part of the waveform at the base. Record this measurement and the other data collected in the following steps in Table 14.2. Notice that this negative exponential voltage at the base holds *Q1* in the cut-off state until the voltage on the left-hand side of *C2*, as it discharges, approaches zero and then *Q1* goes into the conducting state.

2.

Also measure and record the dc voltage at the top and bottom of this exponential trace. You may need to increase the intensity of the beam temporarily in order to see all of the trace at the bottom of the screen. Do not use this high intensity any longer than necessary. It could burn the phosphor off the screen. Note the magnitude and polarity of the voltage on the base (at the bottom of the screen) at the beginning of the exponential curve. How could a negative voltage of this magnitude be present when only a positive dc power source is used for the circuit ?

3.

Measure the dc voltage on the base (with the oscilloscope) during the time that *Q1* is in the saturated state ? Do you notice any little bump at the beginning of this horizontal line? You may need to increase the sensitivity of the vertical volts/div. control in order to observe this anomaly. If one is present then measure its duration and record in Table 14.2. Also measure the dc voltage at the top of this hump. Does this hump go positive at its peak?

Theoretical Calculation of the Duration of the Square Wave at the Collector

Theoretically the duration of the square wave at the collector of *Q1* can be calculated by Eq. (14.1). For *Q1* *R* is *R2* and *C* is *C1*. For the trace at *Q2* then *R* is *R3* and *C* is *C2*:

Table 14.2. Data on the voltage waveform at the base in an astable multivibrator.

Duration of the exponential part of the waveform at the base of Q1	
Maximum negative voltage at the base of Q1 (at the bottom of the exponential)	
Voltage at the top of the exponential	
dc voltage at the base while Q1 was in the saturated state	
Duration of the little bump at the top of the exponential	
dc voltage at the top of this bump	

$$\frac{T}{2} = 0.693 \ RC. \tag{14.1}$$

1.

Record in Table 14.3 the calculated, by the above expression, and the measured duration of one-half of the square wave at the collector of $Q1$.

If $R2$ was not equal to $R3$ or $C1$ not equal to $C2$ then the square waves would not be symmetrical. By using the proper values for R and C it is possible to use this circuit to generate narrow pulses. The author did this in order to get the narrow pulses to use as trigger pulses for the monostable and bistable multivibrators in the text.

Table 14.3. Calculated and measured duration of a half-period of the square wave output of this astable multivibrator.

The calculated duration of one-half of the square wave at the collector of Q1	
The measured duration of one-half of the square wave at the collector of Q1	

15
RC Differentiating Circuits

Components Needed

1. Capacitors: One 0.1 μF and one 470 pF.
2. Resistors: One 4.7 kΩ and/or a resistor substitution box.
3. Diode: One germanium signal diode.

RC Differentiating Circuit

For the circuit in Fig. 15.1 use a 4.7-kΩ resistor or a resistor substitution box for R and a fixed capacitance for C. Do not include the diode at this time.

In the Appendix of the textbook it was shown that an RC circuit, similar to Fig. 15.1 , would differentiate a voltage waveform if the RC time was very short in comparison to the period (T) of the waveform. There are two ways to study the effect of the ratio RC/T on the input waveform. One is to use a waveform with a constant period and then to vary the RC time by using a resistor substitution box to vary the resistance. The other would be to use a fixed resistance and capacitance and then to vary the frequency (f) of the waveform. Since the period equals 1/f this would vary the period of the input waveform.

Both of these methods are included in this manual. It would be repetitive to do both, therefore it is suggested that students do either but not both of these methods. At this time you can continue the following experiment or skip this part and start your experiment on p. 91.

Using a Fixed Frequency but a Variable RC Time

Use a 1-kHz square wave for the input. If a variable frequency source of square waves is not available then use the two-stage JFET amplifier in Experiment 4 on p. 26 of this Laboratory manual. Inject large amplitude sine waves into this amplifier. This should drive the second stage into both saturation and cutoff and thus produce reasonably good square waves whose frequency would be the same as that of the input sine waves. Use a 0.1-μF capacitor for C.

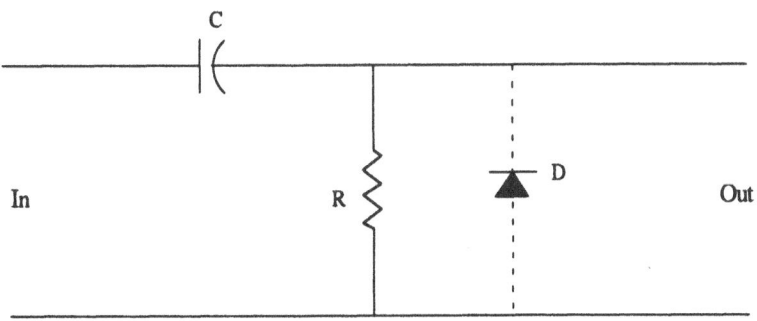

Figure 15.1. An RC differentiating circuit with a limiting diode.

Monitor the input and output waveforms simultaneously with the oscilloscope. Use external trigger obtained at the source of the input waveform.

Effect on the Waveform of the Ratio of RC Time to the Period

Let us first study the effect of the RC time on the output from the above circuit. The period (T) of a 1-kHz waveform would be 1 ms or 1×10^{-3} s. If the resistance was 10 kΩ and the capacitance was 0.1 µF then the RC time would be equal to the period of the waveform. In this study let us vary the resistance from about 100 kΩ which would make the RC time equal to 10 T down to 1 kΩ which would make the RC time equal to 0.1 T.

1.

Use a 1-kHz square wave as the input. Use 0.1 µF for C and 100 kΩ for R. In this case $RC = 10$ T. In Table 15.1 draw the output waveform under the different parts of the square wave as they appear on the oscilloscope screen. Also measure and record the amplitude. Since you may do either part of the experiment the tables are placed at the end of these instructions.

2.

Change R to 10 kΩ and repeat (1) above. In this case $RC = T$.

3.

Change R to 1 kΩ and repeat (1). In this case $RC = 0.1$ T.

4.

If you are using, or have available, a resistance substitution box use it and gradually reduce the resistance from 100 to 1 kΩ while monitoring the output with the oscilloscope. What was the ratio of the RC time to the period (T) when the circuit began to distort the input waveform? What was this ratio when the output appeared to be a differential of the input waveform? At this point sharp vertical lines would appear at the same time as the edges of the square waves. Record this data in Table 15.1.

This circuit is similar to those used to couple two stages of amplification. Why is the waveform not distorted in those amplifiers?

5.

Use 1 kΩ for R so that the output is composed of two narrow pulses, one positive and the other negative. Put the diode in the circuit using the polarity shown in Fig. 15.1. This should cut off almost all of the negative pulse. Measure the amplitude of the positive pulse and the negative stub. Also measure the width or duration of the positive pulse at about the 3/4 maximum amplitude point. Record the measurements in Table 15.2. You might want to reverse the polarity of the diode and observe the results. This is a method of producing sharp trigger pulses of either polarity.

Theory of RC Effect on Voltage Waveforms

Now let us use the function of the components and explain a change in waveform due to a RC circuit. There are two ways to explain how this circuit effects the square wave.

1.

If the RC time is very short, C can charge up fast enough for the voltage across it to follow the input voltage waveform across the circuit. Therefore the only times that charging current would flow through R, and develop a voltage across R, would be when the input voltage was changing. Therefore the output voltage would be zero except during the vertical edges of the square wave. The direction of this current through R during the trailing edge of the positive peak would be opposite to that during the leading edge so the output pulses would have opposite polarities. If the RC time was not short then some charging current would flow through R during at least a part of the positive and negative peaks of the square wave and develop an output voltage.

2.

Mathematically, the differential of a voltage waveform represents the slope of that curve at that instant, or during that short period of time, that the differential is taken. For a square wave the slope would be zero during the positive and negative peaks and maximum (one positive and the other negative) at the transitions between those peaks. Therefore the differential of a square wave would be two sharp pulses, one positive and the other negative, separated by periods of zero volts.

Differentiation of Sine Waves by RC Circuits

While the circuit is performing as a good differentiating circuit switch to a sinusoidal input without changing the frequency. Compare the output with the input with regards to amplitude and phase. Any mathematical handbook would show that if a sine wave is differentiated the result is a cosine wave. If you shift a sine wave 90° to the left it becomes a cosine wave. Is the phase difference between the output and the input about 90° in your circuit? Show this in Table 15.3.

Use of Slopes to Explain Differentiation of Sine Waves

One way of looking at the differentiation of these waveforms is based on the slope at various points on the input waveform. The differential of a curve is often defined as the slope of that curve. Let us examine the input waveform. The maximum slopes in this sine wave occurs at the midpoints between the positive and negative peaks with the positive maximum being at the midpoint preceding the positive peak. The maxima or peaks of the output wave should be in vertical lines with these midpoints of the input wave if the waveform has been differentiated. The slopes are zero at the positive and negative peaks of the input sine wave. Therefore the output wave should pass through its zero point in a vertical line with each peak of the input wave.

Differentiation of Triangular Waves by RC Circuits

If a source of triangular waveforms is available you could switch the input to those waveforms, at the same frequency as for the sine waves, and study the differentiation of these waveforms. A triangular wave is composed of two straight lines, one with a constant positive slope and the other with a constant negative slope. Therefore the differential of a triangular wave should be two constant voltages, one positive and the other negative. This describes a square wave. Show your results in Table 15.4.

Using a Fixed RC and a Variable Frequency

If you have completed the preceding part of this experiment it would not be necessary to perform the part that follows. It would be a repetition of the same study.

Now we will study the effect of the frequency on the shape of the output waveform when the RC time is held constant. For this part of the experiment let us use a capacitance of 0.047 µF and a resistance of 4.7 kΩ. The RC time would be 0.221×10^{-3} s or approximately 0.2 ms. A square wave with this period would have a frequency of 5 kHz. Let us vary the frequency of the square wave from 50 kHz which would have a period (T) of 0.1 RC down to a frequency of 500 Hz which would have a period of 10 RC.

1.

Monitor the input and output of your RC circuit with your oscilloscope. Inject a square wave with a frequency of 500 Hz across the RC circuit in Fig. 15.1. In Table 15.1 draw the output waveform under the square wave as it appears on your oscilloscope screen. In this case $RC = 0.1$ T.

2.

Change the frequency across the RC circuit to 5 kHz and repeat (1) above. In this case $RC = $ T.

3.

Change the frequency to 50 kHz and repeat (1). In this case $RC = 10\,T$.

4.

Monitor the output of the RC circuit with the oscilloscope while you start with a frequency of 50 kHz and gradually reduce it to 500 Hz. Notice the frequency at which distortion first appeared in the output. Also note the maximum frequency that the waveform appeared to be differentiated. This effect is so gradual that it is a matter of judgment. It is usually stated that when RC is 6 or more times T that it is a long time and when RC is 1/6 T or less that it is a short time. Calculate your data to get the ratio of RC to T and record in Table 15.1.

5.

Use 500 Hz so that the output is composed of two narrow pulses, one positive and the other negative. Put the diode in the circuit using the polarity shown in Fig. 15.1. This should cut off almost all of the negative pulse. Measure the amplitude of the positive pulse and the negative stub. Also measure the width or duration of the positive pulse at about the 3/4 maximum amplitude point. Record the measurements in Table 15.2. You might want to reverse the polarity of the diode and observe the results. This is a method of producing sharp trigger pulses of either polarity.

Theory of RC Effect on Voltage Waveforms

Refer to p. 90 for a discussion of this theory.

Differentiation of Sine Waves by RC Circuits

While the circuit is performing as a good differentiating circuit switch to a sinusoidal input without changing the frequency. Compare the output with the input with regards to amplitude and phase. Any mathematical handbook would show that if a sine wave is differentiated the result is a cosine wave. If you shift a sine wave $90°$ to the left it becomes a cosine wave. Is the phase difference between the output and the input about $90°$ in your circuit? Show this in Table 15.3.

Use of Slopes to Explain Differentiation of Sine Waves

Refer to p. 91 for a discussion of this topic.

Differentiation of Triangular Waves by RC Circuits

If a source of triangular waveforms is available you could switch the input to those waveforms, at the same frequency as for the sine waves, and study the differentiation of these waveforms. Refer to p. 91 for a discussion of the differentiation of a triangular wave. Show the effect on a triangular wave in Table 15.4.

Table 15.1. Effect of RC time on square waves.

		Amplitude
Input waveform ►		
Output waveform when RC = 0.1 T		
Output waveform when RC = T		
Output waveform when RC = 10 T		
Ratio of RC to T when distortion first became noticeable		=
Ratio of RC to T when waveform first appeared to be differentiated		=

Table 15.2. Dimensions of pulses resulting from differentiating a square wave.

	Amplitude	Duration at 3/4 Max.
Positive pulse		
Negative stub		

Table 15.3. Differentiation of a sine wave.

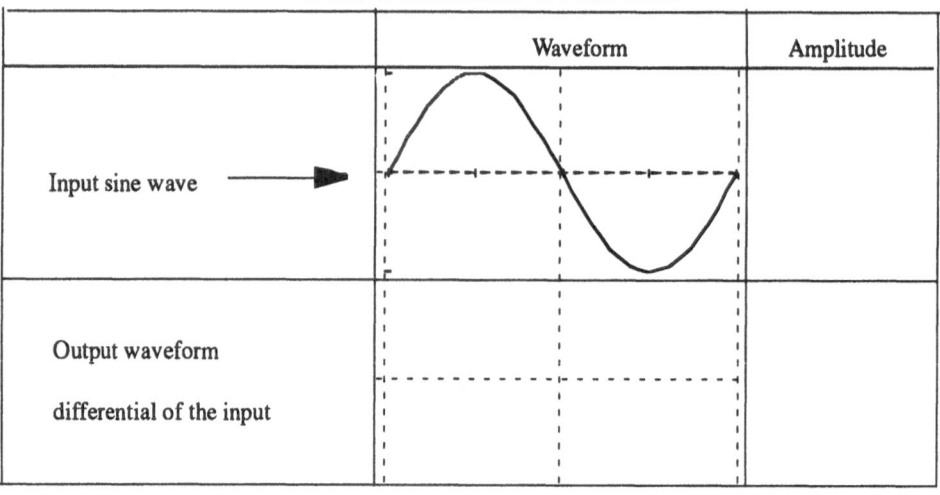

	Waveform	Amplitude
Input sine wave		
Output waveform differential of the input		

Table 15.4. Differentiation of a triangle.

	Waveform	Amplitude
Input triangle wave		
Output waveform differential of the input		

16
RC Integrating Circuits

Components Needed

1. Capacitor: One 0.1 µF.
2. Resistors: One 100 kΩ, one 10 kΩ, and one 1 kΩ. A resistor substitution box can be used instead of these resistors.

RC Integration

If the *RC* time in Fig. 16.1 is short, as compared to the period of the input wave, then the charge on the capacitance, and the voltage across it, could follow the voltage of the input wave. Thus the output, which is across it, would have the same waveform as the input waveform. Therefore with a short *RC* time constant this circuit could be used to couple two stages of amplification. However, it is shown in the Appendix of the textbook that an *RC* circuit, similar to Fig. 16.1, would integrate a voltage waveform if the *RC* time was very long compared to the period of the waveform.

RC Integrating Circuit

Build the circuit shown in Fig. 16.1. See 1 below. Use a 1-kHz square wave as the input waveform. The period of the waveform would be 1 ms. Use a 0.1-µF capacitor as *C*. Observe the input and output waveforms simultaneously with the oscilloscope. Get your external trigger from the source of the input waveform.

RC Integration of a Square Wave

1.

Use discrete resistors with values of 1, 10, and 100 kΩ in that order for *R*. You can use a resistor substitution box to get these values if you prefer. These resistances combined with the 0.1-µF capacitor would have *RC* times of 0.1, 1 and 10 times the period of the input waveform. Draw the output waveforms for each *RC* time in Table 16.1. Record the amplitude also.

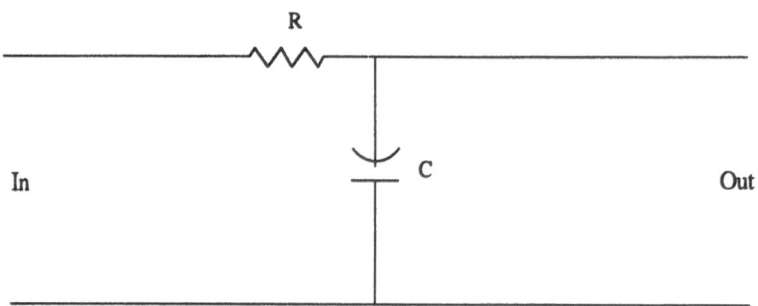

Figure 16.1. A simple RC integrating circuit.

Was the square wave passed undistorted when the *RC* time was short compared to the period of the input waveform? The explanation of the triangular waveforms produced by integrating a square wave follows.

Theory of Integrating Circuit Action on a Square Wave

If the *RC* time is very long compared to the period of the input square wave then the square wave will serve as a dc voltage source to charge up the capacitor. If the *RC* time is long enough, the capacitor will not be fully charged up when the peak of the square wave ends.

In the Appendix it was shown that, when a capacitor is charged by a dc source (V), the voltage (v) at any time (t) can be calculated by the following expression. The fraction v / V represents the fraction of the charging voltage across the capacitance at any time (t):

$$\frac{v}{V} = 1 - e^{-t/RC}.$$

When the charging voltage is a peak of a square wave, (t) would be (T) the period of the square wave. When the 100-kΩ resistor was used t/RC would be 0.1. When these values are substituted in the above expression the result would show that the voltage across the capacitance at the end of the peak would be only about 1/10 of the charging voltage. Actually the voltage across a capacitance while it is being charged by a dc source is an exponential. What you see on the oscilloscope is such a small fraction of a full exponential charge that it appears to be a straight line.

Mathematical Interpretation of Integrating a Square Wave

Mathematically the integral of a curve represents the area between that curve and the zero reference line, between the two points of integration. As you integrate the square wave between the leading edge and the trailing edge of the positive peak the area under the curve would increase in a linear manner. This would produce a triangle with a positive slope. The algebraic expression that would represent the positive peak of the square wave could be stated as v = 0t + k where k is a constant that represents the amplitude of the square wave. An expression that represents the integral of this expression would be v = kt where k would now be the slope of the line representing the voltage as a function of time. Thus the integral of a square wave should be a triangular wave. Does this agree with the result as shown on the oscilloscope? Report your results in Table 16.1 at the end of these instructions.

RC Integration of a Sine Wave

While the circuit is performing as an integrating circuit change the input to a sine wave without changing the frequency.

1.

Compare the input sine wave with the output wave with regards to phase and amplitude. Record the results in Table 16.2. The integral of a sine wave is a negative cosine wave. If your oscilloscope has the capability of inverting one trace then invert the output wave and compare it with the input sine wave to see if the output would then be a cosine wave. On p. 3 of your text it is shown that if a sine wave is shifted to the left 90° it becomes a cosine wave. If your oscilloscope does not have the capability of inverting a waveform you could use the circuit in the introduction of this Laboratory manual to invert the output waveform.

Did you notice that the amplitude of the output wave was much less than that of the input wave when the input was either differentiated or integrated by an RC circuit? This was probably more evident when the input was a sine wave.

RC Integration of a Triangle

If you are using a waveform generator that can put out a triangular wave then use the same frequency that you used for the sine wave as the input to the RC circuit.

1.

Compare the shape of the integral of this triangle waveform to that of the sine wave with the same frequency. See the textbook for a discussion of the integral of a triangle waveform.

Table 16.1. Integration of a square wave.

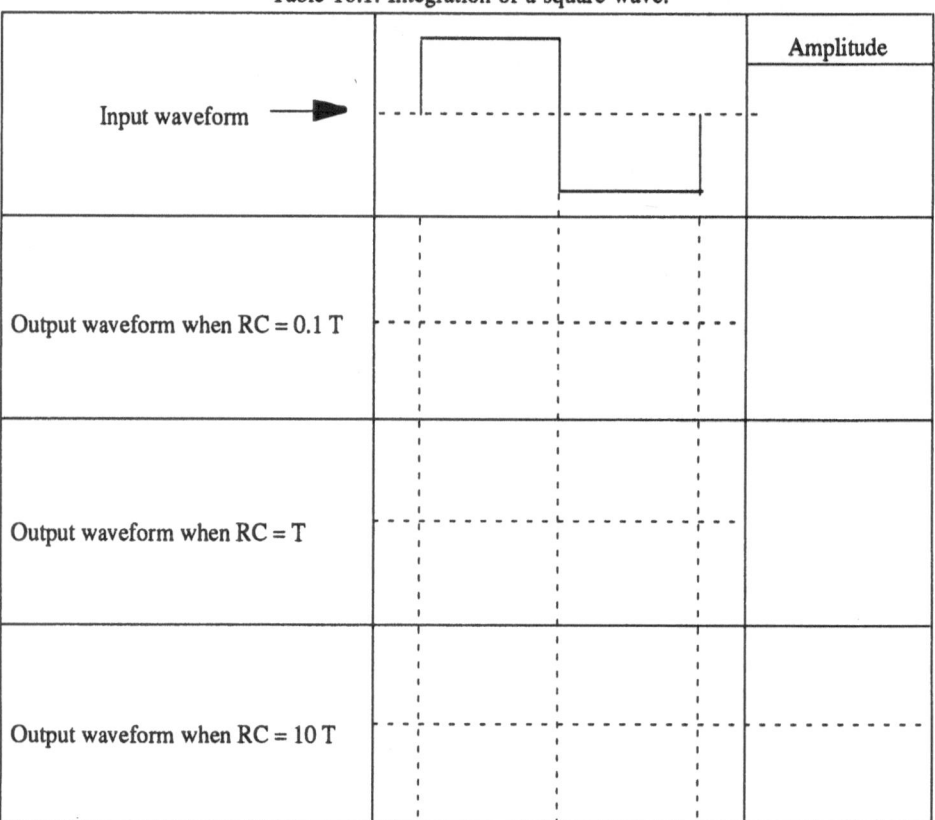

		Amplitude
Input waveform ➤		
Output waveform when RC = 0.1 T		
Output waveform when RC = T		
Output waveform when RC = 10 T		

Table 16.2. Integrating a sine wave.

	Input waveform	Output waveform
Amplitude		
Phase shift	0°	

17
Schmitt Trigger Circuits

Components Needed

1. Transistors: Two *npn* BJTs, for example, 2N2925, MPS2222A, or any general purpose *npn* BJT transistor.
2. Resistors: One 33 kΩ, one 10 kΩ, three 6.8 kΩ, one 4.3 kΩ, and one 330 Ω.
3. Capacitors: One 0.22 μF and one 0.01 μF.
4. Potentiometer: One with any resistance from 10- to 100-kΩ.

Schmitt Trigger Circuit

Build the circuit shown in Fig. 17.1. The values of the components are not critical. The values shown are only to give you the order of magnitude for each one. Use a dc power supply of approximately +10 to +15 volts.

Circuit Study

Do not connect the external oscillator to the circuit until you have tested to see that *Q1* is in the cut-off state and *Q2* is in the conducting state, in the absence of an input trigger voltage. In order to do this, use a high impedance dc voltmeter and measure the dc voltages at the bases, emitters, and collectors of *Q1* and *Q2*. The emitters are connected together so their voltage could be measured at the top of *R6*. The dc voltage at the base of *Q2* should exceed that at the emitters by approximately 0.6 to 0.7 volts. This would indicate that *Q2* was forward-biased. The dc voltage at the base of *Q1* should be zero. Thus *Q1* would be reverse-biased and in the cut-off state. If *Q2* is in the conducting state the dc voltage at the collector should be very slightly greater than the voltage at the emitter. The difference between the voltage at the collector of *Q2* and its emitter should be the voltage drop across a conducting JFET transistor. Notice the amount of this difference. The dc voltage at the collector of *Q1* would not be that of the power supply while in the cut-off state because of the current through the voltage divider composed of *R1*, *R3*, and *R4*. You could calculate the voltage at this collector if you desired and compare that value to the measured value.

Use the oscilloscope to monitor the input (ahead of the coupling capacitor) and the output (collector of *Q2*) while injecting a sine wave, as a trigger for the circuit.

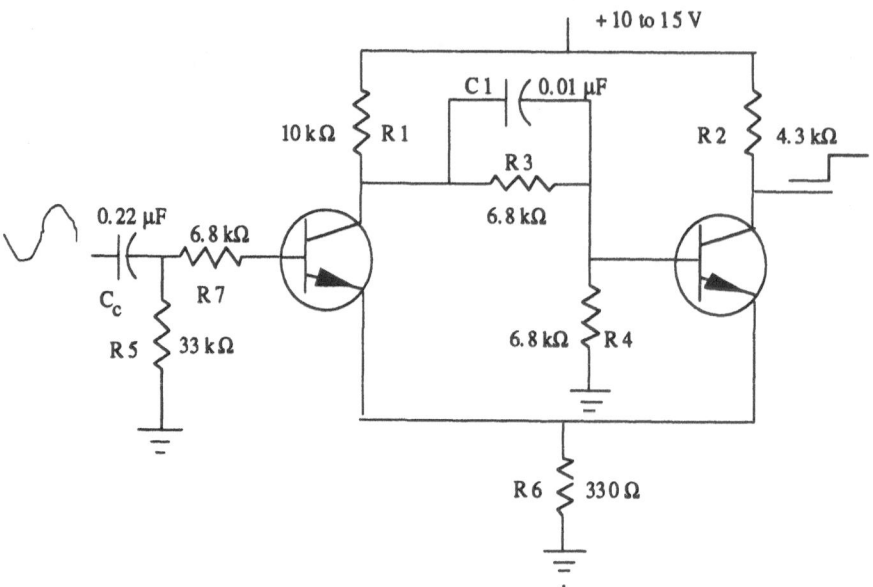

Figure 17.1. Schmitt trigger circuit.

Inject the signal at the input as shown in Fig. 17.1. Use a frequency of 1 kHz for the input sine wave. Any other low frequency could be used. For this part, switch the input of the oscilloscope to ac. Locate the zero volts reference line for the probe that is used to monitor the input sine wave by shorting that probe to ground. Move that zero volts line vertically until it is located on the bottom grid line. Thus you will see only the top half of the input sine wave. Get your external trigger for the oscilloscope at the output of the external sine wave oscillator or at the square wave output of your waveform generator.

1.

Start with a low amplitude of input sine wave and slowly increase its amplitude until a square wave output appears at the collector of $Q2$.

At what voltage on the input sine wave was $Q1$ turned on? In order to determine this, move the trace that shows the output square wave vertically until it just meets the trace of the sine wave input. Then measure from the bottom grid line to the point on the sine wave that $Q2$ is cut off by the trigger voltage. Record these measurements in Table 17.1.

2.

In a like manner determine the voltage point on the sine wave that turns $Q1$ off, and $Q2$ on, and record the result in Table 17.1. Was the amplitude of the input waveform, at the instant that $Q1$ was triggered to the cut-off state, the same as when it was triggered into the conducting state? Can you explain the reason for any difference in these trigger points? What component in the circuit could you change that would affect this hysteresis effect?

Table 17.1. Trigger levels of input sine wave.

	Amplitude of sine wave
Circuit was triggered on at	V
Circuit was triggered off at	V
As the amplitude of the sine wave increased the duration of the output square wave	
As the amplitude of the sine wave increased the amplitude of the output square wave	

Parameters of the Output Square Wave

1.

Vary the amplitude of the input sine wave and note the effect on the duration of the output square wave at the collector of $Q2$. Show this effect in Table 17.1.

2.

Switch the probe that is used to monitor the output square wave to dc and locate the zero volts line on some grid line below the middle of the screen. Measure the dc level of the voltages at the top and bottom of the output square wave and record these values in Table 17.2.

3.

Is the output transistor $Q2$ saturated when it is on? In order to determine this one needs to measure the dc voltage between the emitter and collector when the transistor is in the conducting state. With the inputs of both channels of the oscilloscope switched to dc, locate both zero volt lines on a grid line near the bottom of the screen. Once these zero lines have been determined *do not* move the vertical position control. However you can change the sensitivity (V / div) control.

4.

While the sine wave input is turning the Schmitt trigger circuit on and off, observe simultaneously on the oscilloscope the waveforms at the emitter and collector of $Q2$. Measure, and record in Table 17.3, the difference in dc voltages between the emitter and collector on the oscilloscope when $Q2$ is "turned on." For a silicon transistor to be saturated this voltage between emitter and collector should be approximately 0.3 V.

5.

Using the same technique determine the voltage between the base and emitter of $Q2$ when that transistor is in the conducting state. Record this in Table 17.3.

Table 17.2. Data on the output of the Schmitt trigger circuit.

dc level at the top of the output square wave	V
dc level at the bottom of the output square wave	V
Amplitude of the output square wave	V

Table 17.3. dc voltages at Q2 while in the conducting state.

	dc voltages on Q2 when conducting
Between collector and emitter	V
Between emitter and base	V
Was Q2 saturated when in the conducting state	

Table 17.4. dc emitter voltages

	dc voltage at emitter
Q1 cut off, Q2 conducting	
Q2 cut off, Q1 conducting	

Voltage Waveform at the Emitter

Locate the zero-volt reference line for the probe that is to be used to monitor the output square wave by shorting that probe to ground. Move that zero-volt line vertically until it is located on the third grid line counting from the bottom. Use either ac or dc for this input. For the other probe move its zero-volt line to the grid line just above the middle line on the screen. Switch this probe to dc and observe the waveforms at the emitters as you monitor the output square wave with the other probe.

1.

Measure the voltage at the emitter during the time that $Q1$ is cut off and $Q2$ is in the conducting state and also while $Q1$ is held in the conducting state, and $Q2$ cut off, by the trigger voltage at the base of $Q1$. Record these measurements in Table 17.4. Why is the emitter voltage different when $Q1$ is cut off from that when $Q2$ is cut off? Do you notice an anomaly at the beginning of the bottom line in the waveform at the emitters? For a possible explanation of this anomaly go to p. 163 in your text.

Trigger Levels

Now let us use a different method of determining the magnitude of voltage at the base of $Q1$ that is required to trigger the Schmitt trigger circuit on and off. One method is to use a dc source whose output amplitude can be varied slowly such as a solid state one with a variac in front of it. A second method is to use a 9-V battery across a multiturn potentiometer. Connect this dc source to the input of the Schmitt trigger (not through the capacitor). Monitor this input dc voltage with a high impedance voltmeter and also with one probe of your oscilloscope which is switched to dc. Use the other oscilloscope channel, also on dc, to monitor the output at the collector of $Q2$.

1.

Slowly increase the dc voltage on the base of $Q1$. The pointer on the voltmeter will gradually move upscale. At the same time the horizontal line that is measuring the same voltage at the base of $Q1$ will move upward until the Schmitt trigger circuit is "turned on." At that instant both the pointer on the meter and the trace on the oscilloscope will drop to a lower reading as base current starts in $Q1$. During all of this time the oscilloscope trace, that is recording the voltage at the collector of $Q2$, will remain at a constant level until that instant when the circuit is "triggered on." At that instant $Q2$ would be cut off and this oscilloscope trace would suddenly jump to a higher level. Repeat this procedure a few times. During each trial note the maximum voltage reading on the voltmeter just prior to its sudden drop as the circuit is "turned on." If your voltmeter has been accurately "zeroed" and a sensitive voltmeter scale is used a very accurate measurement of the "trigger on" voltage can be made.

2.

After you have determined the "trigger on" voltage then determine the "trigger off" voltage. Use the same setup as in "1" above. In this case start with the dc voltage at such a level that the Schmitt trigger circuit is "held on" by the dc voltage at the base of $Q1$. Slowly reduce the magnitude of this voltage while observing the voltmeter pointer and the oscilloscope traces. At the "trigger off" level of voltage at the base of $Q1$ the oscilloscope trace at the collector of $Q2$ will suddenly drop to the level normally present when no trigger voltage is present. At the same instant, the voltage at the base of $Q1$ will suddenly rise as the base current stops. This is indicated by the sudden movement of the meter pointer and the trace, both of which are recording the dc voltage at the base of $Q1$. Record both trigger voltages in Table 17.5.

3.

Use this external dc voltage to hold the circuit in the "on" condition and measure the dc voltages at the base, emitter, and collector of both $Q1$ and $Q2$. Use a fine point pen and record these voltages at the appropriate terminals in Fig. 17.1. Remove all triggering voltages and measure the dc voltages at the same terminals and also record either over or under those when the circuit was held "on."

Table 17.5. Trigger voltages when a dc source was used to provide the trigger voltage.

	dc voltage at the base of Q1
Trigger-on voltage	V
Trigger-off voltage	V

18
Integrated Circuit Operational Amplifiers

Components Needed

1. Integrated circuit: One 741 operational amplifier.
2. Resistors: Two 100 kΩ, two 10 kΩ, and four 1 kΩ.
3. Power supply: Two 9-V transistor batteries or any 9- to 15-V dual-polarity power supply.

The name "operational amplifier" was derived from the fact that these circuits could be used to perform mathematical operations on voltage waveforms that described various physical phenomena. Operational amplifiers, made with vacuum tubes and other discrete components, were used many years before transistors and then integrated circuits were invented. These were used as analog computers in the described mathematical operations. The circuit boards for those devices using discrete components occupied about the space that a television circuit occupies at the present time. It was only after the development of the integrated circuit operational amplifiers that their use became common.

This experiment is designed to introduce the students to integrated circuits.

Inverting Operational Amplifier

Figure 18.1 shows the external components that could be used with the integrated circuit to make a typical operational amplifier circuit. Point A is a virtual or phantom ground. The voltage, with respect to ground, at a virtual ground is zero although there is no actual ground connection at that point. These may exist in any circuit that uses both a positive and a negative power supply. These are very common in integrated circuits. In most cases a terminal marked negative must be connected to a negative dc source and not connected to the ground of a positive dc source. In the laboratory a simple dual polarity voltage source can be made by connecting two 9-V transistor batteries in series. The junction between them would be connected to ground in the circuit. In an ideal operational amplifier there would be zero current either in or out of the device at point A. Most IC op amps approach this ideal so this small current at point A can be ignored. If this is true then any signal current that flows through $R1$ must also flow through R_f.

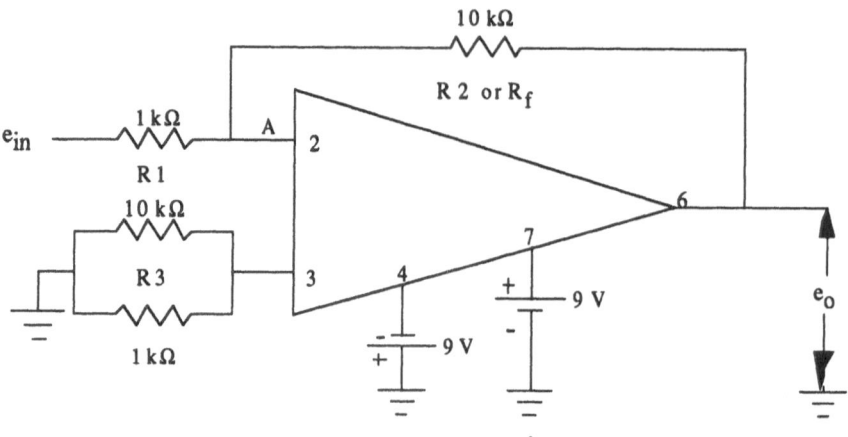

Figure 18.1. Operational amplifier circuit that amplifies and inverts the signal.

The output signal voltage in a 741 op amp is taken between terminal 6 and ground. However, since point A is a virtual ground this same output voltage would be across R_f. Under these conditions the ratio of the amplitude of the output signal voltage, across R_f, to that of the input, across $R1$, would be the same as the ratio of their resistances. Therefore the voltage amplification of this circuit would be the ratio $R_f/R1$ or in this case 10.

Any dc current in or out at points A and B is negligible. However, any imbalance in the dc voltages at A and B would be amplified by the op amp and might cause distortion in the output signal, especially when a large amplification of a signal is desired. Insofar as any input current at point A is concerned R_f and $R1$ are in parallel. Therefore the external resistance at point B should be equal to that of R_f and $R1$ in parallel. In the laboratory one can use two resistors, one equal to $R1$ and the other equal to $R2$, in parallel for $R3$.

Inverting Operational Amplifier Circuit

Build the circuit shown in Fig. 18.1.

Use a 1-kHz sine wave as an input signal. Switch the input of your oscilloscope to ac, and get your external trigger from the output of your signal source. Observe the output signal between terminal 6 and chassis (ground) and the input signal at the left end of $R1$.

1.

Determine the voltage amplification experimentally and compare it to the ratio of $R_f/R1$. We will call this feedback resistor $R2$ instead of R_f in the rest of our discussion. What phase shift occurs from the input to the output signals? Record the results in Table 18.1.

Table 18.1. Data on the inverting operational amplifier.

Resistors used	Phase shift from input to output	Amplitude		Amplification	
		Input	Output	Exp.	R2 / R1
R2 = 10 kΩ, R1 = 1 kΩ					
R2 = 100 kΩ, R1 = 1 kΩ					
R2 = 1 kΩ, R1 = 1 kΩ					

2.

Replace *R2* with a 100-kΩ resistor and the 10-kΩ resistor in *R3* with a 100-kΩ resistor and repeat the above.

3.

Use a dc voltmeter or the oscilloscope on dc and see if there is any dc voltage component in the output signal.

4.

Change *R2* from 100 to 1 kΩ and the top resistor in *R3* to 1 kΩ and repeat (1) above. This makes an amplifier that should invert a signal without changing its amplitude.

Noninverting Amplifier Circuit

Replace the resistors with those shown in Fig. 18.2. Change the ground (chassis) connection from the point in Fig. 18.1 to the left-hand side of *R1* and inject the signal at the left-hand side of *R3* which was shown as ground in Fig. 18.1. This reversal of connections will make the noninverting amplifier in Fig. 18.2.

1.

Repeat (1) and (2) above. In this configuration the voltage amplification will be the ratio of *R2/R3*. If you substitute the values for *R3* in the above ratio it will show that the amplification of this noninverting amplifier is one more than that of the inverting amplifier when the same values are used for the resistances. Record your data in Table 18.2.

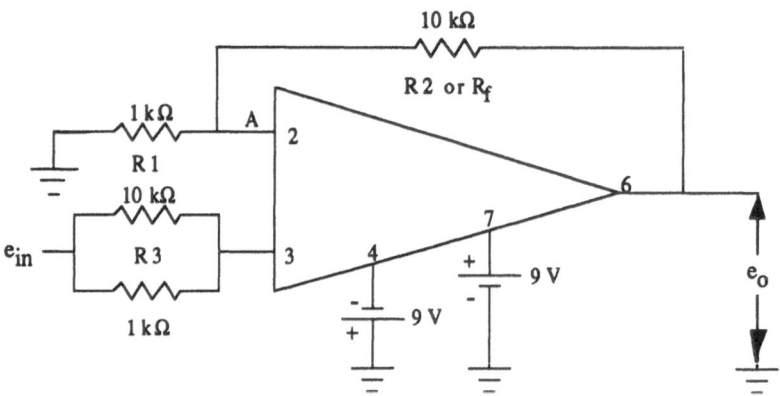

Figure 18.2. A noninverting operational amplifier.

Table 18.2. Data on the noninverting operational amplifier.

Resistors used	Phase shift from input to output	Amplitude		Amplification	
		Input	Output	Exp.	R2 / R3
R2 = 10 kΩ R3 = 10 kΩ and 1 kΩ in parallel					
R2 = 100 kΩ R3 = 100 kΩ and 1 kΩ in parallel					

19
Operational Amplifiers as Differentiators and Integrators

Components Needed

1. Integrated circuit: One 741 or other operational amplifier.
2. Resistors: One 100 kΩ, two 10 kΩ, and one 270 Ω.
3. Capacitors: One 0.1 μF and one 470 pF.
4. Voltage sources: Two 9-V transistor batteries.

When the proper external components are used, an operational amplifier can act as a differentiator of voltage waveforms. When different external components are used, it can act as an integrator. Both of these cases will be studied in this experiment. In experiment 15 starting on p. 93 and in experiment 16 on p. 98 you were asked to draw the resulting waveforms when square waves, sine waves, and possibly triangular waves were differentiated. Therefore, in this exercise you can refer back to those drawings instead of drawing them again. Since the invert mode of the op amp is used, you will need to invert the output to get the true differential or integral. Do this with the oscilloscope or use an inverting op amp with a gain of one.

Operational Amplifier Differentiating Circuit

Build the circuit as shown in Fig. 19.1. Use a frequency of 1 kHz as the input signal. Switch the input of your oscilloscope to "ac" and get external trigger at the source of the input signal. Observe the input and output waveforms simultaneously.

1.

Use a square wave with a frequency of 1 kHz as the input waveform. Measure and record the amplitudes of the input and output waveforms in Table 19.1. Compare the amplitudes and the shapes of these waveforms with those when a simple RC circuit was used in the previous experiment.

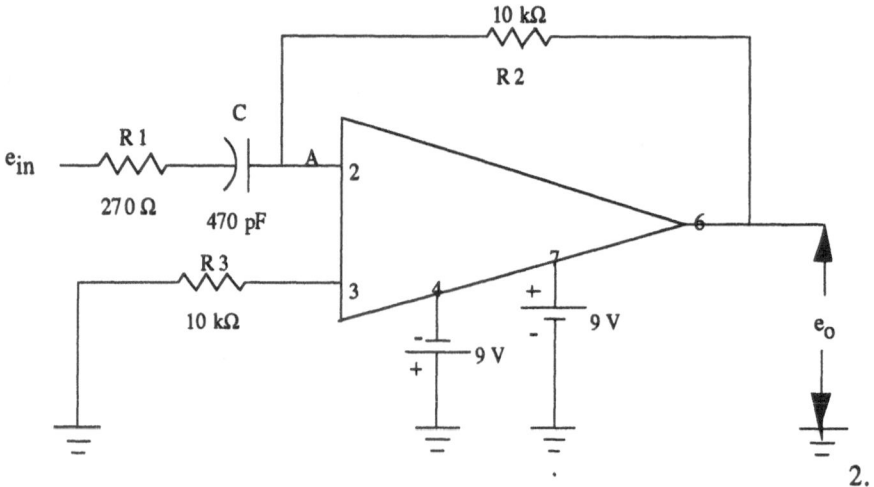

Figure 19.1. Operational amplifier differentiating circuit.

Table 19.1. Data on the differentiation of waveforms by an operational amplifier.

Waveform	Frequency	T / 2	Operational amplifier			Simple RC circuit		
			Amplitudes		Amplifi-cation	Amplitudes		Amplifi-cation
			Input	Output		Input	Output	
Square wave								
Sine wave								
Triangle								

2.

Repeat the above when the input is a sine wave.

3.

If you are using a waveform generator as the source of your input waveforms then repeat the above using a triangular wave as the input waveform.

Operational Amplifier Integrating Circuit

Change the circuit into an integrating circuit by using the configuration shown in Fig. 19.2. Use a frequency of 1 kHz again.

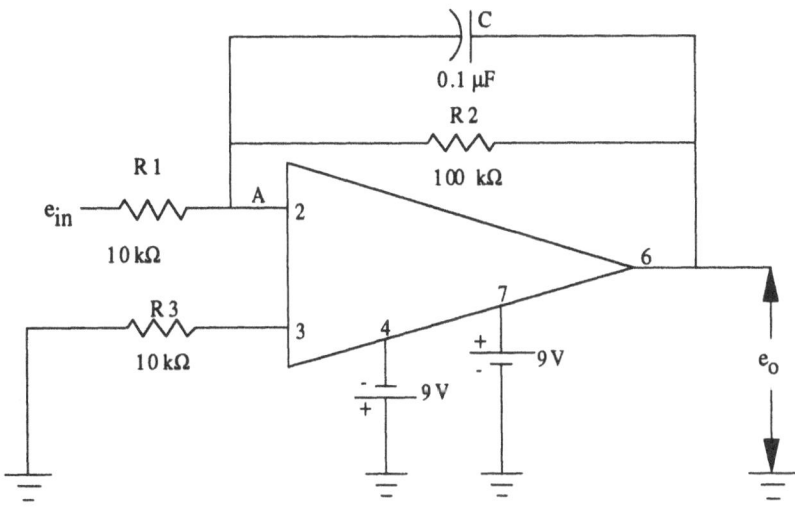

Figure 19.2. Operational amplifier integrating circuit.

Table 19.2. Data on the integration of waveforms by an operational amplifier.

Waveform	Frequency	T / 2	Operational amplifier Amplitudes Input	Operational amplifier Amplitudes Output	Amplifi-cation	Simple RC circuit Amplitudes Input	Simple RC circuit Amplitudes Output	Amplifi-cation
Square wave								
Sine wave								
Triangle								

1.

Use a square wave as the input waveform. Measure and record the amplitudes of the input and output waveforms in Table 19.2. Compare these waveforms with those when a simple RC circuit was used in the previous experiment.

2.

Repeat the above using a sine wave, with the same frequency, as the input waveform.

3.

Repeat 2 above using a triangular waveform if you have a function generator.
    It is easy to predict the shape of the output waveform as a result of the differentiation when the input is a square wave or triangular wave by using the idea that the differential represents the slope of the curve. It is also easy to predict the shape of the output when a sine wave is either differentiated or integrated, by using the information in tables of differentials or integrals.

It is also easy to predict the shape of the output wave when a square wave is integrated by using the idea that the integral represents the area between a curve and the reference line. You can see that the area changes in a constant manner as you trace across a positive or negative peak.

It is more difficult to predict the result of integrating a triangular wave. Probably the best way to predict this result is to use a Fourier Series to represent a triangular wave as follows:

$$v = \frac{8V}{\pi^2} \left[\sin x - \frac{1}{9} \sin 3x + \frac{1}{25} \sin 5x - \frac{1}{49} \sin 7x + \cdots\cdots\cdots + \frac{1}{n^2} \sin nx \right].$$

Notice that the amplitudes of the 3rd, 5th, and 7th harmonics, as indicated by their coefficients 1 / 9, 1 / 25, and 1 / 49 are so low compared to the fundamental (sin x) that the triangular wave approaches a sine wave. Therefore the integral of a triangular wave should approach the integral of a sine wave, which is the negative cosine. If you did integrate a triangular wave, you probably noticed that the output resembled that which resulted from the integration of a sine wave. You might want to refer to the photograph of an integrated triangular wave in the textbook.

20
Waveform Generators

Components Needed

1. Integrated circuit: One ICL 8038 sometimes listed as 8038 CCPD, LM 566 CN, or some other waveform generator.
2. Resistors: One 82 kΩ, one 22 kΩ, and two 1.5 kΩ.
3. Capacitor: One 0.033 µF.
4. Power supply: 10 to 30 V positive polarity.

Circuit Using a 8038 I C

Figure 20.1 shows a Function generator IC 8038 along with the necessary external components needed to function as a waveform generator.

The values for the external components are not critical. Note that this IC requires only a positive-polarity power supply. It is possible to use a dual-polarity power supply of plus and minus 15 V. The only difference would be that the resulting voltage waveforms would move symmetrically about ground if a dual-polarity power supply was used. When a single-polarity (positive) is used the entire waveform is positive with respect to ground or chassis. The magnitude of this voltage determines the maximum amplitudes of the output waveforms. This function generator puts out sine waves, square waves, and triangular waves. In order to vary the amplitude of these waveforms a multiturn potentiometer could be used across the output.

The frequency or repetition rate can be selected externally over a range from 1 Hz to 1 MHz. When the two resistances at terminals 4 and 5 are equal then the output frequency can be calculated by the following expression:

$$f = \frac{0.3}{RC} .$$ (20.1)

In this expression R is the resistance used at pin 4 or 5 and C is the capacitance between pins 10 and 11. Therefore a wide range of output frequencies can be produced by using different values for R and C. However, there is one constraint on the values that can be used.

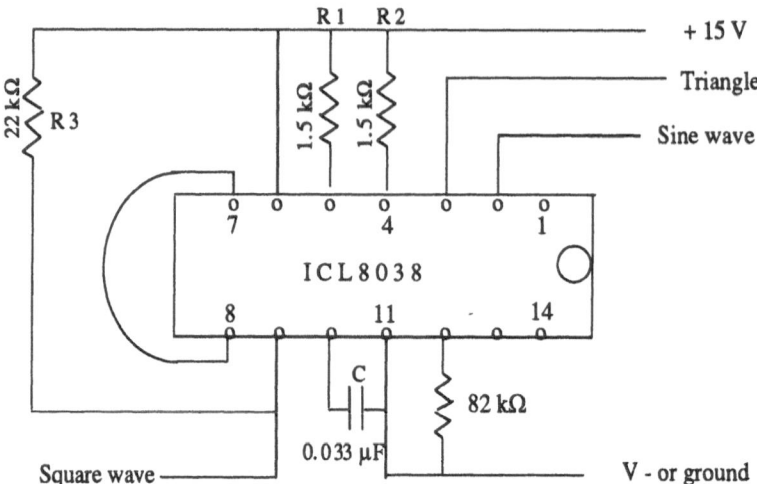

Figure 20.1. Function generator IC 8038 along with the necessary external components needed to function as a waveform generator.

It is the "charging current" which can be calculated by the following equation:

$$i = \frac{V}{5R} \cdot \qquad (20.2)$$

In this equation V is the power supply voltage and R is the resistance at pin 4. This current must not be lower than 1 µA or greater than 5 mA.

Sometimes it is better to use a variable resistor or a resistor substitution box between pins 11 and 12 instead of the 82-kΩ resistor shown in Fig. 20.1. This resistance can then be varied in order to improve the shape of the sine wave output if that is a problem.

Function Generator Circuit

Build the circuit as shown in Fig. 20.1. Use internal trigger for the oscilloscope until you see that there is a square wave output at pin 9. If a square wave is present at pin 9 then get external trigger at that terminal for the rest of the experiment.

1.

Observe and measure the amplitude and period of the square wave, the sine wave and the triangular wave outputs. Calculate the predicted frequency and the period using Eq. (20.1). Measure the power supply voltage and then calculate the charging current using Eq. (20.2). Record these data in Table 20.1.

2.

Change the two resistances $R1$ and $R2$ at pins 5 and 4 to 15 kΩ and repeat (1) above. Record your data in Table 20.2.

3.

Use $C = 0.1$ µF and $R1$ and $R2 = 15$ Ωk and repeat (1). Record your data in Table 20.3.

Frequency Modulating the Output Sine Wave

The frequency of this waveform generator is a function of the dc voltage at pin 8. If a sinusoidal voltage is applied at that terminal the output sine wave of the generator would be frequency modulated.

1.

Use the sine wave oscillator, or function generator, that you have been using in the previous experiments to inject a low frequency sine wave between pin 8 and ground or V- of the 8038 IC. Use a coupling capacitor between your external oscillator and pin 8 so you will not short out the dc voltage at pin 8. In this way you will be producing a variation in the dc voltage that would be applied at pin 8. Use the sine wave voltage of this external oscillator as the external trigger voltage for your oscilloscope as you monitor the sine wave output of the 8038 IC. It may take a fine adjustment of this trigger voltage and the frequency of the external oscillator to produce a stable FM waveform.

Table 20.1. Data on the function generator using the given values of R and C.

	Pin 4 or 5	Between pins 10 and 11	Period			Charging current
Waveform	R	C	Calculated	Experimental	Amplitude	Calculated (i)
Square	1.5 kΩ	0.033 µF				
Sine	1.5 kΩ	0.033 µF				
Triangle	1.5 kΩ	0.033 µF				

Table 20.2. Data on the function generator after changing R to 15 kΩ .

	Pin 4 or 5	Between pins 10 and 11	Period			Charging current
Waveform	R	C	Calculated	Experimental	Amplitude	Calculated (i)
Square	15 kΩ	0.033 µF				
Sine	15 kΩ	0.033 µF				
Triangle	15 kΩ	0.033 µF				

Table 20.3. Data on the function generator when $C = 0.1\ \mu F$ and $R = 15\ k\Omega$.

Waveform	Pin 4 or 5 R	Between pins 10 and 11 C	Period Calculated	Experimental	Amplitude	Charging current Calculated (i)
Square	15 kΩ	0.1 μF				
Sine	15 kΩ	0.1 μF				
Triangle	15 kΩ	0.1 μF				

21
Audio Power Amplifier

Components Needed

1. Integrated circuit: One IC 380, 383, 386, or 1877, or any other integrated circuit audio power amplifier.
2. Resistors: Depends on amplifier selected. None for the 380 IC.
3. Capacitors: One 0.47 µF and one large capacitance, for example, 100 µF, or greater, are recommended for the 380 IC.
4. Power supply: 6 to 24 V (positive polarity). Two 9-V transistor batteries, connected in series, could be used instead of this power supply.
5. Loud speaker: A small one made for transistorized circuits. A headphone could be used.

Amplifier Circuit

This experiment is designed to give the student an opportunity to work with a consumer product and "have fun." The IC selected was the LM380. The only external components needed for laboratory use of this device are two capacitors. One of these might not be necessary if the audio input to pin 2 has no dc component. Use a power supply of 15 to 20 V with a positive polarity. The voltage amplification of the 380 is fixed internally at 34 dB. The numerical voltage gain could be calculated from the following expression:

$$dB = 20 \log \frac{V \text{ (out)}}{V \text{ (in)}} \text{ or } dB = 20 \times \log \text{ of the voltage gain,}$$

$$34 = 20 \log A_V.$$

The voltage gain A_V would be the inverse log of 1.7 or 50. Therefore the voltage gain would be fixed internally at 50 for the 380 IC audio voltage amplifier.

If you are using a different integrated circuit amplifier, the external circuit is usually included along with the specifications sheet that you should request when you purchase the IC.

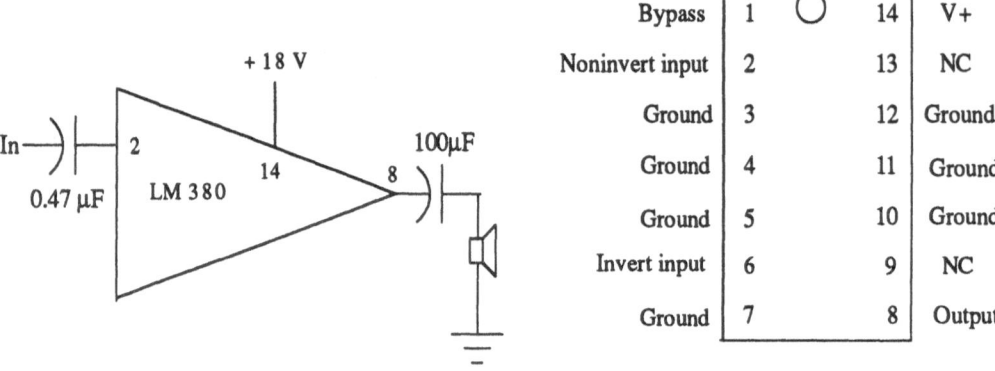

Bypass	1	○	14	V+
Noninvert input	2		13	NC
Ground	3		12	Ground
Ground	4		11	Ground
Ground	5		10	Ground
Invert input	6		9	NC
Ground	7		8	Output

Figure 21.1. Circuit for laboratory use of an LM 380 power amplifier.

Circuit Construction

Build the circuit as shown in Fig. 21.1. A capacitance of some value would be needed to block any dc component at pin 8. A large capacitance is needed at this output in order to transfer most of the output signal voltage to the loudspeaker and not lose it across the capacitance. If you do not have a capacitance as large as the size recommended then use the largest capacitance that is available.

Inject a 1-kHz signal at the input, pin 2. Use an amplitude that does not distort the output signal. Get external trigger at the output of the audio source. If the output is distorted you might try placing a 0.47-µF capacitor between pin 1 and ground. Observe the input at pin 2 and the output at pin 8.

1.

Measure the amplitude of the signal at the input and output and calculate the voltage gain. Compare the phase of the output with that of the input. Record in Table 21.1.

2.

Use a high impedance voltmeter, or the oscilloscope (on dc), and measure the dc component of the voltage at the output (pin 8). Record in Table 21.1.

Using the LM 380 IC to Amplify Music

Replace the external oscillator with a source of music or other sound. Probably the best source of an input audio signal would be a tape recorder and recorded music. If there is no dc component in the input audio signal $C1$ could be either included or eliminated. You could determine this in the same way that you did in (2) above.

1.

Observe the music on the oscilloscope and listen to it at the loudspeaker at pin 8. You may need to mute the output of the speaker of the tape recorder. Is the sound as clear at the output of the 380 as at the output of the tape recorder or is it "mushy?"

2.

You could use a potentiometer between the output of the source of music and the input to the amplifier and use it as a volume control if you desired. You could also use a variable resistance in series with a capacitance and make a tone control if you desire. Both of these are shown in Fig. 21.2.

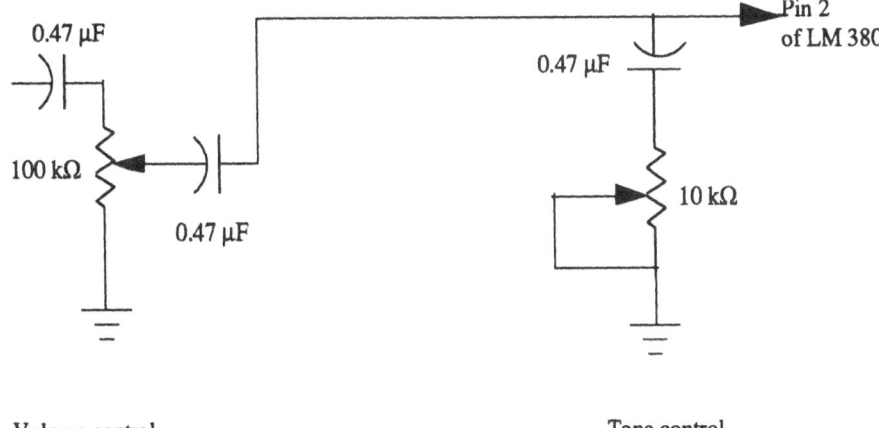

Volume control   Tone control

Figure 21.2. External volume and tone controls for an audio amplifier.

Table 21.1. Data on the LM 380 power amplifier.

| | Amplitude | | Amplification | Phase shift |
	Input	Output	Experimental	Input to output
1-kHz sine wave				
What was the dc voltage at the output (pin 8)?				
Was the music distorted?				

22
The Hall Effect

Components Needed

1. One Hall probe.
2. Amplifier: One voltage amplifier, for example, 741 IC or other integrated circuit operational amplifier.
3. Resistors: Two 2 kΩ, two 100 kΩ, and one 1 kΩ.
4. Potentiometer: One 10 kΩ.
5. Power sources: Two 9-V transistor batteries, one 6-V battery, and a low voltage (variable) dc source (see p. 122). You can use a dc power supply with a variable voltage ac transformer.
6. Milliammeter: One that can be used at about 200 mA.

Theory of the Hall Effect

When a slab of conductor or semiconductor material, carrying an electric current, is oriented properly in a magnetic field a potential difference is developed across two sides of the slab. This is the Hall effect. For a discussion of the theory of the Hall Effect refer to p. 379 in the textbook. This effect can be described by Eq. (Z2) on p. 382 which is the following:

$$R_H = \frac{V_H d}{B I} .$$

This can be changed to the following expression:

$$V_H = \frac{R_H B I}{d} .$$

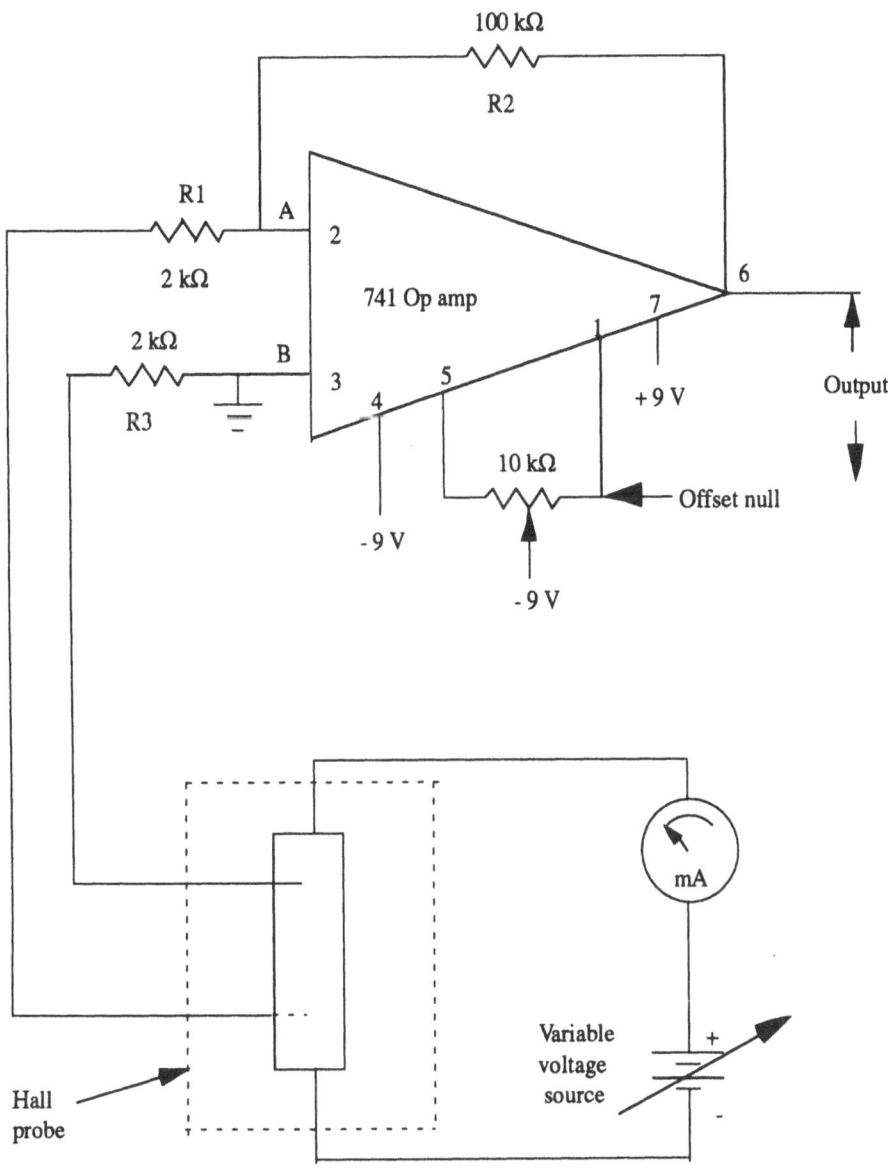

Figure 22.1. Setup for Hall voltage measurements.

In this expression R_H is the Hall coefficient for the particular Hall device and is determined by the amount of the dopant element put in the semiconductor when the device was made. The factor d represents the thickness of the device and is measured in meters. These two factors are constant for a particular device and are usually stated in the specifications that are furnished with the device.

The current I, in this expression, is measured in amperes and is held constant when using the device. Under these conditions the Hall voltage V_H would then be proportional to B which is the strength of the magnetic field in Gauss.

In this experiment we will keep the current through the device constant and then use a magnet with a known magnetic flux at its surface to calibrate the setup in gauss per Hall volt. If a calibrated magnet is not available then data can be collected when a tiny magnet is placed against the side of the Hall device. Then the Hall coefficient and the thickness can be used along with your current and Hall voltage measurements to calculate the magnetic field associated with your tiny magnet. Then you can use it as your calibrated magnet.

Then the Hall probe can be used to measure the strength of other magnetic fields. Since the Hall voltage is low in magnitude we will amplify it so that it can be measured in our experiment. It is necessary that we know the amplification factor of the amplifier so that we can calculate the actual Hall voltage. We will use an operational amplifier for this purpose. A typical setup for this experiment is shown on the preceding page.

The Hall device that you use may be different from the one that is being described. In that case follow the instructions that accompany the device that you are going to use. Be very careful when working with Hall devices. They are very thin and very fragile.

The Hall Probe Circuit

Connect the operational amplifier as shown in Fig. 22.1 and then check its operation before connecting it to the Hall probe. You can check it by placing a sinusoidal signal from an oscillator or function generator at the left end of *R1* with the common or ground lead at *B*. Use an oscilloscope to see if the circuit is amplifying properly. Since we are going to use a dc current through the Hall probe, the Hall voltage that we want to measure is dc. Therefore we will not determine the voltage gain until later.

1.

Use a digital meter or FET meter to measure any dc voltage that might be present at the output of amplifier circuit. If any is present then adjust the offset null to reduce it as much as possible.

2.

Connect the two inner leads of the Hall device to the inputs of the operational amplifier. Connect the two outer leads in series with a variable dc current source and a milliammeter. The current source, Hall probe, and the milliammeter must be connected in series and proper polarities must be observed. If you do not have a source of variable dc current you can use a low voltage dc voltage source which is "plugged in" a variac and then use the variac to control the current which would be measured by the milliammeter. This current must not exceed that recommended in the Spec. sheet. A suggested value for the first measurement would be 100 mA. Do not exceed 200 mA in any part of this experiment.

3.

Increase the input voltage source across the Hall probe while watching the milliammeter. If no deflection appears then turn off the power and double check your circuit.

4.

With a known constant current through the Hall probe but no magnet near the probe, measure the dc voltage at the output of the amplifier with your high impedance voltmeter, for example, FET. Then adjust the 10-kΩ offset potentiometer until the output measures zero volts on the lowest range on your meter. If you cannot zero the output with the offset potentiometer then try reducing the current through the Hall probe. Sometimes it is necessary to disconnect the leads of the Hall probe from points A and B and then short A and B together and adjust the offset potentiometer for zero output. This may be due to a small magnetic field in the room near the probe. If you do not quite get the output to zero but it is very small compared to one volt then proceed with the experiment.

   If all attempts at balancing the op amp to near zero offset fails then you could get a variable low voltage ac source and an ac milliammeter and pass ac current through the Hall probe instead of dc. You would then need to replace the dc meter at the output of the op amp with an oscilloscope in order to measure the Hall voltage.

5.

With a constant dc or ac current through the Hall probe, place the small calibration magnet against the Hall probe so that the Hall voltage as measured at the output of the operational amplifier is maximum. With the amplifier output voltage, as registered by the meter, and the known magnetic field of your calibration magnet you can calibrate the amplifier output in gauss per volt.

6.

Now place the Hall probe in an unknown magnetic field. *Hold the probe firmly* as you gently move the magnet up to it. From the calibration in gauss/volt and the amplifier output voltage, when the probe is in the unknown magnetic field, determine the strength of the magnetic strength in gauss.

7.

For strong magnetic fields use a low current through the Hall probe and for weak magnetic fields use a larger current but never exceed 200 mA or the maximum as stated on the Spec. sheet.

IMPORTANT

You must recalibrate your setup in gauss/volt for each different current through the Hall probe.

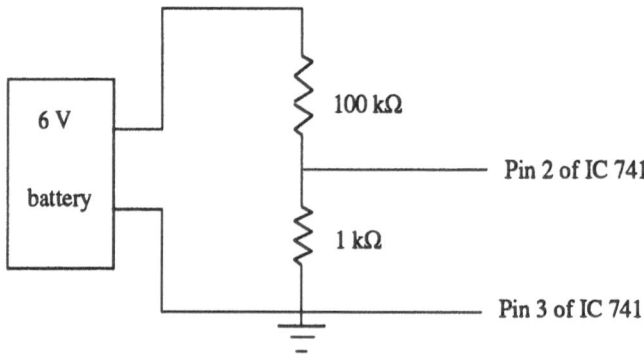

Figure 22.2. Circuit used to determine the dc voltage amplification of our amplifier.

8.

In order to solve for R_H you must determine the voltage gain of your amplifier so that you can find the actual Hall voltage developed by the probe. In order to do this use the setup shown in Fig. 22.2. The ratio of $R2$ to $R1$ in our amplifier circuit is 50 so theoretically the voltage amplification should be 50. However this factor is so important in this experiment that we need to determine the measured amplification.

With this setup the input voltage would be about 0.06 V so the output of the operational amplifier should be about 3 V for our setup. Measure the input and output voltage amplitudes and determine the actual dc voltage amplification. Divide your measured Hall voltage by the amplification factor to find the actual Hall voltage to record in Table 22.1.

Studies Using the Hall Probe Circuit in Figure 22.1

1.

Using a known magnetic flux determine the Hall coefficient (R_H) using a few different values of current (I). Record in Table 22.1. Determine the mean value of R_H and compare with the value as given on the Spec. sheet for your Hall Probe. If you do not have a known magnetic field then place a small magnet against the Hall probe. Measure the Hall voltage and use the value of R_H for your Hall probe and calculate the magnetic field of your small magnet. Then use it as your known magnetic flux.

2.

Using different values of current (I) for calibration make a few measurements of the magnetic flux (B) for an unknown magnet. Determine the mean value of (B) for that magnet. Record your results in Table 22.2.

Table 22.1. Experimental values for the parameters in Eq. (22.1).

Current (I)	Hall voltage	d	B	Calculated value of R_H
Mean experimental value of R_H				
R_H value on the Spec. sheet				

Table 22.2. Experimental parameters used to find the magnetic field of our unknown magnet.

Current (I)	Hall voltage	Gauss/volt	Calculated value of B in Gauss
Mean experimental value of B for the magnet			
Actual value of B (if known)			

23
Standing Waves On Twin-Lead Transmission Lines

Material Needed

1. Radio frequency generator with a range up to at least 50 MHz.
2. Electrical wire: Number 12 or 14 gauge house wire long enough to reach across the laboratory.
3. Voltmeter: One high impedance analog meter with a dc voltage range down to a fraction of a volt, for example, an FET meter.
4. Diode: One germanium signal diode.
5. Resistor: One 470 Ω if using number 14 wire; 430 Ω if using number 12 wire.
6. Bolts and nuts: Number 32 size; four bolts, eight washers, four nuts, and four screw eyes.
7. Scrap lumber: A short piece of 1 × 4 board.

Figure 23.1 is a photograph of components to be made for this experiment.

Preparation of the Components

Make two saw kerfs exactly 4 cm apart and about 1/4 inch deep lengthwise in the 1 × 4 board. Then cut off small sections similar to items *A* and *B* in Fig. 23.1. Cut off enough of these sections to maintain a parallel pair of wires at a uniform distance apart (4 cm). Cut two rectangular pieces of wood about 7 cm wide and 9 cm long similar to *E* in Fig. 23.1. Put two screw eyes exactly 4 cm apart in one end of each. These are to be clamped at each end of the laboratory desk or table top with C clamps to provide a little tension on the parallel wires in order to maintain the 4-cm distance between them during the experiment.

Item *C* in Fig. 1 is to be used to detect the voltage between the wires as it is moved down the parallel pair of wires. It should be cut about 3.3 cm wide and 7 cm long. Drill 1/8-inch holes about 5 cm apart on each side of *C* and insert a size 6-32 bolt with nut and two washers in each hole, as shown in the figure. Secure a germanium diode (*D* in Fig. 23.1) between these washers on one side of *C* and a piece of bare wire on the other side. The orientation of the diode is not important. Its function is to rectify the rf voltage so that a dc meter, with its low range, can be used to measure the voltage between the lines at points along the wires.

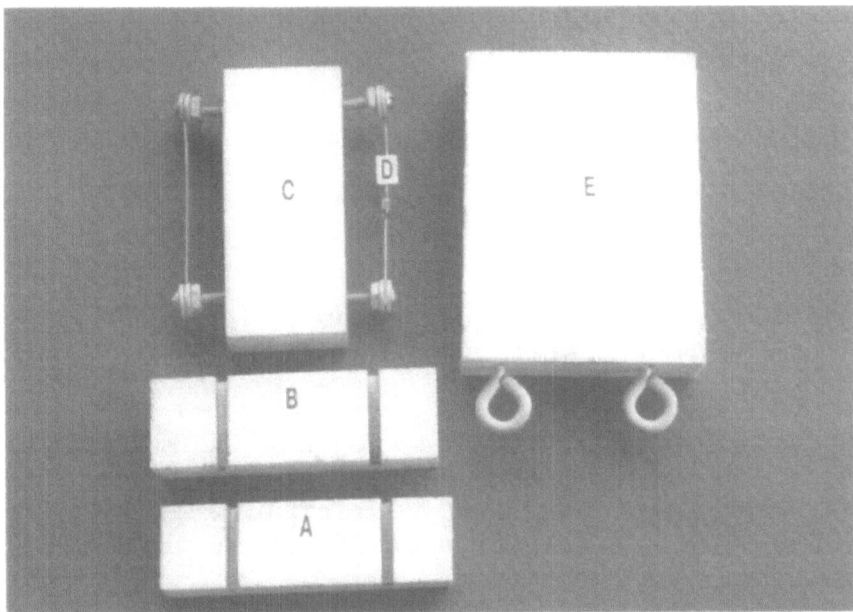

Figure 23.1. Photograph of the components to be used in this experiment.

Strip the insulation from two of the number 12 or 14 house wires. Use a length that will be sufficient to make your twin-lead transmission line. Make your transmission line as long as possible in the area you have. Clamp the two pieces (*E* in Fig. 23.1) to the table top or other surface securely with C clamps and stretch the two wires between the "eyes" in the two pieces. Then position the sections (*A* and *B* in Fig. 23.1) along the line so that the wires are maintained at a uniform 4 cm apart.

Standing Wave Experiment

Connect the output of your rf generator to the two wires at one of the pieces (*E*). Adjust the output of the generator to its maximum frequency and maximum amplitude. You might need to use an oscilloscope or some other means to determine the frequency of the rf oscillator.

Calculate the characteristic impedance using the following expression:

$$Z = 276 \log \frac{2D}{d} .$$

In this expression D is the distance between the two wires, center to center, and d is the diameter of each wire.

For our experiment we made D = 4 cm. The diameter d of number 12 wire is 80.81 mils or 0.205 cm and the diameter of number 14 wire is 64.08 mils or 0.163 cm. Let us assume that you are using number 14 wire. The above expression would then be the following:

$$Z = 276 \log \frac{2 \times 4}{0.163} \ \Omega.$$

$$Z = 466.7 \ \Omega.$$

Standing Waves Along an Open Transmission Line

1.

Adjust your analog meter to its most sensitive dc voltage range. Clip the leads of your meter to the small bolts at one end of the piece marked C in Fig. 23.1. In order to use the meter slide the two bolts at the other end of C, along the two wires and observe the deflection of the pointer on the meter. If the pointer moves downscale then reverse the connections of your meter leads at the small bolts on C to cause an upscale movement on the meter.

2.

With the terminal end of your transmission line open, place the detector (item C) across the line at its terminal end. If the voltmeter leads are oriented correctly the meter deflection should be at least midscale. Record this meter reading at the open end of the transmission line in Table 23.1. Slide the detector (item C) along the line toward the source of rf (the generator). Notice that the amplitude varies in a sinusoidal manner as the detector moves along the transmission line. Record in Table 23.1 the meter reading and the location (distance from the terminal end) at each maximum and minimum.

3.

The standing wave ratio (SWR) is the ratio of the maximum amplitude to the minimum amplitude at adjacent peaks and nodes. Calculate the standing wave ratio near the terminal end of the transmission line and record in Table 23.1.

Table 23.1. Data on an open transmission line.

	Maxima			Minima			SWR
	1	2	3	1	2	3	
Distance from terminal end							
Amplitude							

Table 23.2. Data on a shorted transmission line.

	Maxima			Minima			SWR
	1	2	3	1	2	3	
Distance from terminal end							
Amplitude							

Standing Waves Along a Shorted Transmission Line

1.

Short across the terminal end of the transmission line with a piece of copper wire or a nail. Starting at the terminal end, slide the detector along the two wires toward the source of rf. Record in Table 23.2, the amplitudes of the meter readings and the locations of the maxima and minima along the line. With the transmission line terminated with zero resistance the locations of the maxima and minima should be reversed from those when the output was open or terminated with an infinite resistance.

2.

The rf electromagnetic energy is transmitted along an open lead transmission line with approximately the speed of light or about 3×10^8 meters per second. Calculate the wavelength for this energy using the expression (velocity = frequency × wavelength) for the frequency of your rf generator. In Table 23.3, compare the distances between maxima and the distances between minima with a 1/2 wavelength, and the distances from the terminal end to the first node (for open) and first maximum (for shorted) to 1/4 of a wavelength.

Table 23.3. Calculated and experimental standing wavelengths on open and shorted lines.

Calculated 1/2 wavelength	cm
Distance between adjacent maxima	cm
Distance between adjacent minima	cm
Calculated 1/4 wavelength	cm
Distance from open terminal end to first minimum	cm
Distance from shorted terminal end to first maximum	cm

These are standing waves and are the result of the combination of the incident energy traveling down the line from the generator and the energy reflected back from the load. If a transmission line is not terminated in its characteristic impedance, energy will be reflected back toward the source. This reflected energy will combine with the incident energy and standing waves will be produced.

Result When the Transmission Line is Terminated in its Characteristic Impedance

Select a resistor as near to the calculated characteristic impedance as possible, for example, 470 Ω if using number 14 wire, and lay it across the wires at the terminal end of the transmission line. The nearest value to 439 Ω would probably be 430 Ω if using number 12 wire. This would approximately terminate the transmission line in its characteristic impedance. Under these conditions there should be no reflected energy and thus no standing waves along the line.

1.

Take the detector (item C) and slide it along the two wires from the terminal end toward the source of rf at the generator. There should be a deflection on the meter that is almost constant as the detector moves along the line. The deflection might be slightly greater at the source end than at the terminal end due to some energy loss as it moves down the line. Record the meter readings at both ends of the transmission line in Table 23.4.

Table 23.4. Data on a transmission line when terminated with its characteristic impedance.

Calculated characteristic impedance	Ω
Resistance placed across terminal end	Ω
Meter reading at source end	V
Meter reading at terminal end	V